Michael J. Yochim FOREWORD BY WILLIAM R. LOWRY

REQUIEM FOR AMERICA'S BEST IDEA

National Parks in the Era of Climate Change

High Road Books Albuquerque

HIGH ROAD BOOKS is an imprint of the Unversity of New Mexico Press

Printed in the United States of America

Library of Congress Cataloging-in-Publication Data

Names: Yochim, Michael J., author. | Lowry, William R. (William Robert), 1953– writer of foreword.
Title: Requiem for America's best idea: national parks in the era of climate change / Michael J. Yochim; foreword by William R. Lowry.
Description: Albuquerque: High Road Books, 2022. | Includes bibliographical references and index.
Identifiers: LCCN 2021039330 (print) | LCCN 2021039331 (e-book) | ISBN 9780826363435 (cloth) | ISBN 9780826363442 (e-book)
Subjects: LCSH: National parks and reserves—Environmental aspects—United States. | Forests and forestry—Climatic factors—United States.
Classification: LCC SB482.A4 Y63 2022 (print) | LCC SB482.A4 (e-book) | DDC 333.78/30973—dc23
LC record available at https://lccn.loc.gov/2021039330
LC e-book record available at https://lccn.loc.gov/2021039331

Founded in 1889, the University of New Mexico sits on the traditional homelands of the Pueblo of Sandia. The original peoples of New Mexico—Pueblo, Navajo, and Apache—since time immemorial have deep connections to the land and have made significant contributions to the broader community statewide. We honor the land itself and those who remain stewards of this land throughout the generations and also acknowledge our committed relationship to Indigenous peoples. We gratefully recognize our history.

COVER ILLUSTRATION View of mountains and lake at Glacier National Park in Montana © Stella Levi | istockphoto.com
DESIGNED BY Mindy Basinger Hill
COMPOSED IN 11/15pt Adobe Calson Pro

DEDICATED TO THE MEMORY OF *Michael J. Yochim*,

BELOVED SON, BROTHER, UNCLE, FRIEND, AND

INSPIRATION TO MANY

CONTENTS

MAPS

FOREWORD *William R. Lowry*

I first got to know Mike Yochim back in the 1990s, when I was doing research on national parks and he was working for the National Park Service (NPS) in Yellowstone. When we met, he was helpful and friendly, encouraging of my project. I also learned that his family lived just outside of St. Louis, so our communications and interactions increased over time. Whenever he was back in St. Louis, we would get together, usually for lunch, although he also occasionally stayed with me and my wife, Lynn. Whenever I was in Yellowstone, or later in Yosemite when he worked there, I would visit and sometimes stay with him or catch a meal at some local place like K-Bar Pizza in Gardiner, Montana. We had several things in common in addition to our midwestern roots, the most important being our love of natural places, especially the national parks.

By saying that, I don't mean to suggest that I knew the parks nearly as well as Mike did. Yes, I worked in them, visit them when I can, and have enjoyed some of the experiences Mike discusses in this book, such as rafting the Grand Canyon. But Mike, he lived them. He worked in Yellowstone for twenty-two years, Yosemite for another five, Sequoia for two seasons, and Grand Canyon for one off-season. He was a planner for the NPS, and quite good at it, to the point where he served as the deputy project manager for completion of the *Merced Wild and Scenic River Management Plan* and the project manager for the *Tuolumne River Management Plan* in Yosemite in the early 2010s. But the parks were more than a vocation for Mike—they were an avocation. He was the epitome of what the famous author Edward Abbey said all wilderness advocates should be: He did what he could for these wonderful places, but he also savored any chance to enjoy them. He was a legendary hiker, an avid backpacker, and a world-class cross-country skier. His wilderness companions also swear that Mike was a gourmet backcountry cook.

So you can only imagine our shock when he flew into St. Louis in 2013 and told Lynn and me that he had ALS. At the time, we knew little about the disease and reacted slowly to his disclosure. But he knew enough about it to start sobbing as he said something I will never forget. "I wish I had cancer," he told us. "If I had cancer, they would at least know what to do." I, we, have all since learned a lot about ALS (amyotrophic lateral sclerosis), more than we ever wanted. In short, as Mike told us, few victims of ALS survive even five years. And survival is nearly all that most ALS patients can hope for, as it destroys the ability to control one's entire body.

The disease intensified. I saw Mike half a year after his 2013 visit to St. Louis at his home in Gardiner. Before he was diagnosed, he and I had agreed to cowrite a paper for a conference in Bozeman. Mike's health had already started to decline by then, so I stopped by his house to see if he could come with me, and met his folks, Jim and Jeanne, for the first time. Mike was in no shape to go to the conference, so I delivered the paper without him and we subsequently worked on it until we got it published.[1] Mike moved back to the house near St. Louis, where he grew up, so his parents could take care of him. Lynn and I started to visit on a regular basis and do what we could to help. We went on a few hikes with them and took Mike on a few with just us. But much as Mike loved getting out into the natural areas of Missouri, it got harder and harder to do so. Within a couple of years, Mike, the man who led so many others on hikes into the most remote parts of the country, couldn't even stand on his own. He then used a wheelchair to go on hikes, but even those adventures reminded us of how challenging life can be for disabled people, especially those with ALS.

Ultimately, Mike could barely move at all on his own. He lost the ability to swallow or drink. His parents had to perform all self-care activities for him. I got a sense of their sacrifice one spring when Jeanne had surgery and I assisted by going over to their house early in the mornings to help get Mike up and dressed. He always wanted to look presentable for any visitors, asked to keep having a shower every other day, and, using his eyes, picked out his clothes for the day. That's just one example of how he never gave in to the disease. Neither did his parents. And they never complained or made him feel bad about their sacrifice. Not once.

Over the last few years of his life, the only thing Mike could control were his eyes. So he used them. He wrote. Boy, did he write. Mike wrote two other books and nearly finished this one, all after getting the disease. Using just his eyes and an eye-tracking machine, he would pick out letters to make words and sentences. He would sit at his computer screen every morning and afternoon, with a nap in between, and write about the parks and the forces, like climate change, that are affecting them. This entire book was written with his eyes. It was exhausting work and took him full hours to write a single paragraph.

Why did he do it? Because he cared. Like I said, he loved the parks—and he was doing what he could for them, in gratitude for what they did for him. Mike conveys those memories in his writing with such beauty that it will make just about anybody want to go for a park adventure. At times, his writing makes me think of John Muir, since, in at least some ways, Mike was to Yellowstone what Muir was to Yosemite. Just as Muir embraced every part of Yosemite, Mike embraced every part of Yellowstone. One of his claims to fame is that he hiked all 1,200 miles of trails in the park; he knew it intimately. At other times, his writing is like that of the aforementioned Edward Abbey, full of passion for the natural world and anger at those who are abusing it. And finally, this last book of Mike's reminds me of Norman Maclean, who was trying desperately to stay alive long enough to understand and write about what happened during a forest fire at Mann Gulch, Montana, in 1949—and produce the legendary *Young Men and Fire*.

Mike also used his eyes in conversations, particularly when he told a joke. That may sound strange, but he never lost his sense of humor. His eye-tracking machine, like the one Stephen Hawking used, allowed him to type in words that could then be spoken. And although he couldn't smile, his eyes seemed to do just that whenever he made a wisecrack or teased one of us. Sometimes he had his lines already keyed up in the computer and would inject them at opportune moments in the conversation. For example, we had this running joke he made just about every time we visited. He would tell us what he had been writing lately and then ask me pointedly, "So, my fine, tenured, full professor friend, what have you been working on?" Whatever work I had been doing seemed trivial compared to his efforts. Until the very end, he could always make others laugh, even when his active mind was trapped inside a body that was deteriorating. I have never seen such courage before and likely won't again. I can only hope that I show nearly as much in any challenge I ever face.

Mike died on February 29, 2020. The fact that it was a leap day did not surprise us—a rare day for a rare person. I was honored to serve as a pallbearer at Mike's funeral, along with his brothers and a nephew and a cousin. Indeed, my wife and I had become very close with his family, especially his parents. They are strong and stoic people, always ready with a smile, but the funeral was tough on all of us, especially when his brothers delivered heartbreakingly poignant eulogies that revealed just how much they and we had lost. We were reminded of Mike's code, the advice he gave to himself shortly after getting the disease. He wrote it out on a piece of paper, stuck it to his refrigerator, and used it to finish one of his books, *A Week in Yellowstone's Thorofare*.

> Seek wildness and beauty
> Enjoy the love of family and friends
> Find joy in the here and now
> Be strong, live well, and love deeply

When Mike died, he was working at his desk in the afternoon, like he did most days. His parents noticed that he was writing an email to me, perhaps with a gentle reminder that he had asked me, in the event of his death, to complete this manuscript and turn it into a book. I had of course agreed to do so. Not only could I not say no to Mike, but as I said, I too love the parks and want people to understand the important things happening to them that Mike describes in this book. In addition, I had indeed learned about ALS and want others to do so as well. I worked with Mike's brother Brian Yochim and Mike's friend Eric Compas—we made some edits, compiled the photos, wrote some captions, created some maps, and added this foreword, but nearly all the words you will read are Mike's.

It was an honor to be part of this work. Mike was, simply put, the most inspirational person I've ever known. He dealt with his fate with grace, humor, and a concern for things larger than himself that was truly remarkable. All three of those qualities are evident even in the passage that follows. As far as I know, it was the last one he ever wrote.

> I enjoy being outside, sitting and reading in the extensive gardens my folks have around their house. The property next door is thickly wooded, so we see deer on a regular basis, and an occasional red fox or raccoon. Birds like it here too, finding feeders and houses scattered about. One afternoon, I was sitting beside a birdhouse that was busy with house wrens bringing food to their chicks. I thought I saw one chick fledge, and soon its nestmate did as well—right onto my armrest! Its maiden flight wasn't long and came with a rough landing, for it was on its side with its claw around my finger. It righted itself and then looked right at me, then flew away. Such a neat encounter with nature's wildness, right in the backyard. Will we act in time to give future generations the same experiences?

Requiem for America's Best Idea

INTRODUCTION

If nature's musicians are birds, the canyon wren is the premier soloist of the Colorado Plateau, that deeply dissected area of southern Utah and its neighboring states. Such a simple, yet evocative birdsong, composed of six or eight descending trills or whistles, without the warbling that some of its cousins inject into their melodies. The plaintive song seems to float down from the heavens, as if it were the auditory accompaniment to the grand visual symphony everywhere on display there. This three-inch-tall musician, the color of the dusky red sandstone prevalent in the region, is seldom seen but regularly heard in southern Utah canyon country. True to its name, the wren prefers to live in canyons, building its nest in small recesses in the canyon walls and eating insects it finds in cracks and crevices in those walls. The bird has never been observed to drink water, getting what it needs from the bugs it eats.[1] It's well adapted to its environment, ranging throughout the American West, almost anywhere there are canyon walls. However, I've rarely encountered it outside of the Colorado Plateau, so the canyon wren's lilt evokes the beauty and feel of the Southwest for me. With its descending notes, the song could be sad to some, but with its melodious simplicity, it's also a celebration of the plateau's haunting beauty, a lyrical welcome to the wren's world of sandstone and cliff.

One of the more memorable times I heard the wren's lilt was in 2013, on a weeklong hike through the Narrows of the Escalante River in Utah. There the canyon walls close in, enveloping the river—thought to be the last one explored and mapped by contemporary Americans—in five-hundred-foot-tall walls of red sandstone. The river itself pales in size to rivers almost anywhere else, barely deserving the same moniker we give to the Colorado or the Mississippi, just ten or twenty feet across and six inches or a foot deep. But such terms are relative to the others around them, and the Escalante is the biggest show around, so it's a "river" in a Utah sense of fulfillment. Boots and kayaks are the means of exploration today, neither one perfectly suited to the task: the one gets trashed walking in and across the river, the other grounds out in the ever-present shallows. The wren, of course, doesn't concern itself with such matters, choosing the airways above the bipedal mammals and their transportation travails.

But the punishments are worth the rewards as the river leads the explorers downstream, deeper into the wren's country. Cottonwood trees, shrubby willows, and sticklike horsetails line the stream, gracing it with textures and color. Smooth cottonwood leaves flutter in the afternoon breeze, and chest-high willows bend permanently downstream as if recalling the floods of yesteryear. The knee-high horsetails that sequester silica in their cell walls provided frontier settlers with a different name for the abrasive stalks—scouring rush. As this landscape embodies, water is the font of life in the desert: step away from the river and you'll find Gambel oak and juniper; step farther away, dryland grass and prickly pear. Above the greenery rises the ubiquitous red sandstone, a contrast to the linear oasis it enfolds. Here are sheer cliffs streaked with dark-gray iron and manganese oxides—known as desert varnish—deposited by rainwater dissolving the mineral out from layers above and leaving it where it evaporated. Around the bend are huge, overhanging alcoves formed by rocks falling away along concave jointing in the bedrock. Towering above it all are waves and beaches of rock, frozen in sandstone and time. It's a landscape like no other in North America; indeed, the canyon wren is quite discriminating in its choice of landscapes to serenade.

On that trip in 2013, the wren's song seemed more restive than celebratory, as I told my hiking companion Josh on the second-to-last night. I had been noticing odd changes in my body for the previous two or three months, like seemingly endless muscle twitching, slurred speech when I was tired, and an enhanced tendency to cry. Still subtle, these symptoms weren't noticeable to others, as Josh soon confirmed. But they were worrisome enough to send me to my doctor, especially after I plugged the two most pronounced symptoms into a Google search and got a long and exclusive list of websites about ALS—amyotrophic lateral sclerosis. This is probably *the* most feared neurological disorder, because there is no cure or even effective way to slow the disease progression (at the time; there is a treatment now that slows the rate of decline by about 33 percent). Instead, the disease's victims see their muscles gradually weaken and waste away as the neuronal connections linking them to the brain are incrementally destroyed. Sufferers gradually become imprisoned in their own bodies, their muscles becoming too weak for the person to care for themselves, eat, drink, speak, or even breathe. Eighty percent die within five years of diagnosis, usually of respiratory arrest. But, in my case and at that point in time, I had just begun the process of determining the cause, and the doctors I'd seen assured me that it was probably something different, something more common and treatable. They referred me to a neurologist, who I would not see for another month. So Josh added his reassurance to the doctors', and our conversation moved on to other things. In the back of my mind, though, the Google search lingered, never really going away. As we continued our journey the next day, I found I could only really suppress it with the thought that, whatever was responsible for the changes I was experiencing, I would be glad I had danced once more to the song of the canyon wren.

The next day, we came across a sign that the future of the Colorado Plateau could be as uncertain and tumultuous as my own. Like my symptoms, its decline was subtle, just a few horizontal lines on a rock wall where Coyote Gulch joins the Escalante. They are high-water marks from Lake Powell, from 1999, the last time it was full. Before the Glen Canyon Dam was built, the Escalante flowed into the Colorado River, which today supplies most of the water to the lake. Since 1999, the Colorado's flows have been anemic, the river's headwaters beset by the same long-term drought affecting the Colorado Plateau when we did that hike. Lake levels have fallen accordingly, bottoming out in 2005 at 145 feet below full pool. They have recovered some since then, but the highest they've been since 1999 is about 60 feet below full.[2] When Josh and I turned to go up Coyote Gulch, the lake level was almost 100 feet below full, and there was no sign the lake had ever been seen there other than the high-water marks. Riverside vegetation was undisturbed and cottonwood trees were full grown, both suggesting the lake had been out of sight for a long time, enough for the vegetation to get reestablished from the high lake levels and corresponding inundation of the 1980s and 1990s. Long-term droughts like the one gripping the region at the time are nothing new for the plateau, though, so it might be a while before the lake is back to full, with consequent impacts on the vegetation at that confluence.

Ordinarily, the climate here would switch to a wetter regime after a few years, or even decades, of drought (megadroughts lasting fifty to a hundred years, or more, can be found throughout the archeological record there), but this time could be different. Like most of the American West, the Colorado Plateau is warming up due to climate change. Already, the plateau has warmed about 2 degrees Fahrenheit, with the first decade of the twenty-first century being the warmest in the record books, and the era since 1950 being the warmest in the last six hundred years. By the end of this century, another 5.5 to 9.5 degrees Fahrenheit of warming is expected, assuming the currently increasing rate of CO_2 production continues (which seems a reasonable assumption, given humanity's inability to meaningfully address this threat).[3] Precipitation changes are harder to predict, but it is certain that there will be less available moisture due to the evaporative effects of such warmer temperatures. Under these conditions, the current drought is likely to become the new norm for the region—or a pleasant memory, as global warming takes the region into uncharted climate terrain.[4] It's unlikely Lake Powell will ever fill again, and small streams like the Escalante may dry up completely, with associated effects on the nearby riparian vegetation. If the canyon wren could understand the changes coming to its lovely home, it would surely sing a funeral dirge—or move to cooler climes. Whether the bird will persist as its native climate unravels is unclear.

The Escalante, here part of Glen Canyon National Recreation Area, is part of our national park system, managed by the National Park Service, which oversees most of the country's national park units.[5] Escalante is hardly alone in seeing its climate beginning to change, or in facing massive changes to that climate in the future. Just about any of the country's western national parks is already seeing warming occur. Glacier National Park has lost over 80 percent of its glaciers since park establishment in 1910, while glaciers in North Cascades National Park in Washington now occupy less than half the area they covered in 1900. National parks in general are warming almost twice as much as the rest of the country, 1.3 degrees Fahrenheit per century, compared to 0.7 degrees for the country as a whole.[6] The country's other wild places, especially its wilderness areas (federal lands protected in a roadless, natural state in perpetuity by the Wilderness Act, often on US Forest Service lands) are seeing the same impacts. Sometimes the evidence is readily discernible, such as wildfire and its increasing frequency and severity, while other times it is more subtle, such as the failure of a songbird to take up its annual summer residency. Whatever the effect, the National Park Service (NPS), US Forest Service (USFS), and Bureau of Land Management (BLM) are concerned about the future impacts to the lands and treasures they protect. The list of resources at risk of damage by climate change is long indeed, encompassing just about anything the agencies safeguard. Instead of providing an overview of them here—who could stomach such a list?—this book will take an in-depth look at a few of the most iconic parks and some of the resources at risk in them. That's the first of three major purposes of the book you're holding: to examine the climatic, ecological, and hydrological changes the parks are already seeing, along with the projected changes in those same attributes for later this century.

Just as wild places throughout the West are being affected by climate change, there are invalids everywhere being affected by debilitating illness. Close to 40 percent of us will be diagnosed with cancer in his or her lifetime, and 22 percent of us will die of it. Heart disease, the most common affliction in America today, kills one in three of us.[7] Every one of the millions of us who have a life-threatening disease has a story to tell of our experience with the illness, of its onset, of our day-to-day struggles to deal with it and perhaps overcome it. These are stories of the human experience, stories that evoke emotion and community in those who read or listen to them. They inspire and motivate us, giving us strength to face our own challenges and adversities. This book is the story of my own struggle with terminal illness, a journey that took me away from my adopted home—Yellowstone National Park, and more recently, Yosemite, both of which employed me as a park ranger—and back to my native land, Missouri. It's a story with many parallels to the escalating climate change in the national parks. Both are stories of urgency: just as I don't have much time left, we don't have much time left to protect the parks from the worst of climate change. But they are also stories of contrast, for as much as my affliction is of unknown origin, the other is very much self-inflicted. Ultimately, both are stories of loss: the one inexplicable and inexorable, the other unnecessary and still preventable. This story is the book's second purpose.

Radical change links the parks, threatened by global warming, to the human experience, threatened by terminal disease, but there is something far more powerful joining the two. It's the power of inspiration that wild places possess, merely by virtue of their existence. This attribute of parks and wilderness areas has long been known; indeed, it's one of the reasons they were established in the first place. Some of the national parks were, in fact, used by the military for this express purpose during World War II.[8] But it bears repeating that, in this era when people in government are often indifferent or even hostile to environmental causes, those of us facing life-threatening illness need wild places for respite, inspiration, and imagination. Even if we can't visit wild places, even if we have never been to the parks, even if we don't have pictures of the wilderness areas, wild places inspire and confer peace of mind. Just knowing they exist brings calm—the knowledge that somewhere there are beautiful places where the shackles of disease do not bind, where we can be free to roam and explore. They provide grist for the imagination, motivation to hope, and food for the soul. In a word, they provide resilience. Exploring some of the inspirational meanings wild places have for me, and by extension everyone, is the book's third and final purpose.

The book has five chapters, each focused on one national park, the phase of my disease progression, and climate change discussions that are embedded within a long backcountry trip I took in each park. Parks include Olympic, Grand Canyon, Glacier, Yellowstone, and the Sierra Nevada national parks in California (Yosemite, Sequoia, and Kings Canyon, combined due to their similar resources). All chapters consider what unbridled climate change will do to the parks under discussion by the year 2100—a time well within our grandchildren's lifetimes. All of the chapters chronicle my search for comfort and inspiration as disease overtook me. The climate change discussions are based on the most recent information available, as found in over two hundred published research papers and books (almost all of them published since 2000), reputable internet sources such as national park websites and a nonpartisan climate modeling website, materials provided to me by park managers, and personal knowledge and experience. In some cases, the narrative looks to historical events to provide a glimpse of a possible future for a park. The book is not a dry, academic tome, but rather a conversation about the effects of climate change, present and future, on a few western national parks, presented within an extended visit to each park. This approach brings to life the tremendous resources preserved in the parks, resources never more threatened than they are now. In short, there are many books on climate change and many on national parks, but few on the intersection of these two topics, and none that presents their confluence in this manner—and that also discuss the importance of the parks to disabled people.

Celebrated western writer Wallace Stegner once proclaimed, "National parks are the best idea we ever had. Absolutely American, absolutely democratic, they reflect us at our best rather than our worst." Yet today, a radically different force, also of our own making, is threatening that noble idea, changing the parks as irrevocably as terminal disease changes me. Glaciers are melting, forests are burning, and animals are disappearing from habitats they once found adequate for their needs. To prevent the worst from happening—extinctions, ecosystem collapse, continuous summertime heat wave and wildfire—the book closes with a call to action, demanding change from our elected officials, our church leaders, and ourselves. Ultimately, if the book is an elegy for our premier national parks, it's also a plea to take action now. Only in so doing will our children and grandchildren be able to experience these fantastic places as I have been fortunate to do, to create their own cherished memories, to celebrate wild places and the life they impart to us. The canyon wren's future is in our hands. Will future generations be swooned by its evocative lilt? Will our actions reflect us at our best—or our worst, as it currently appears?

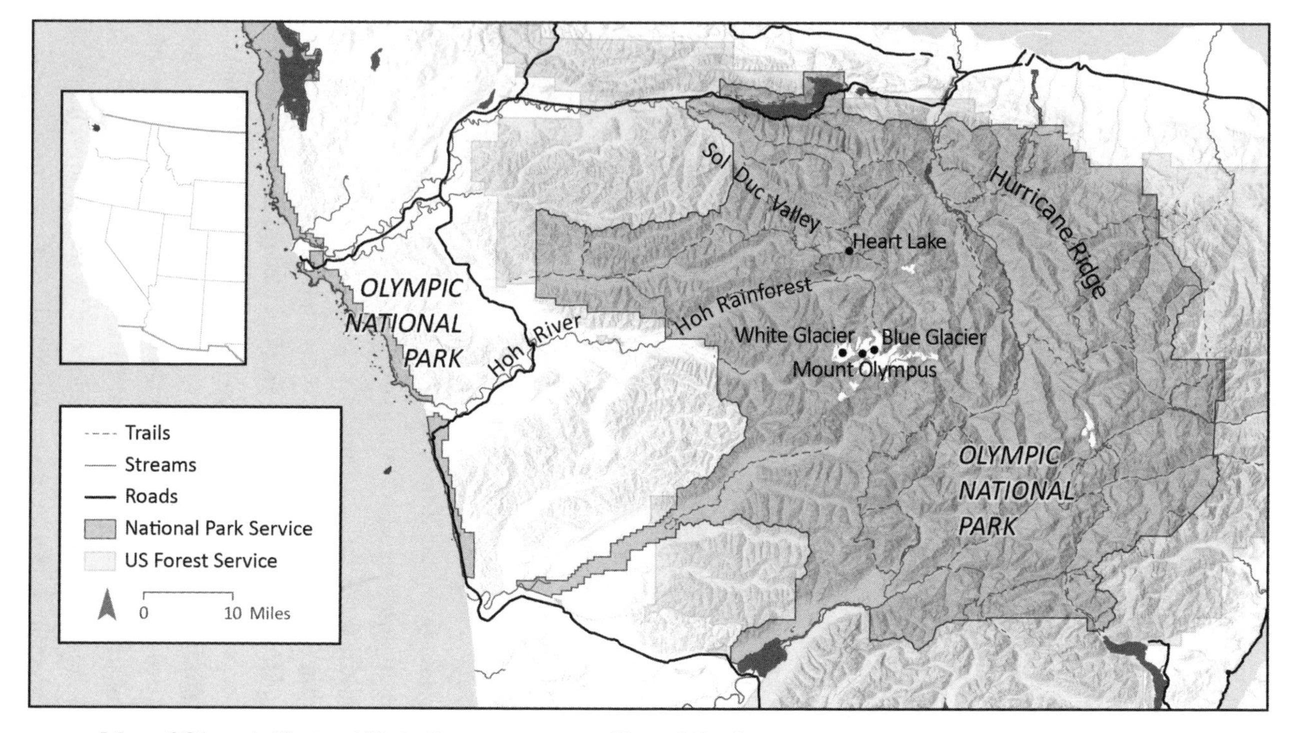

MAP 1 Map of Olympic National Park. *Sources*: NPS, USGS, Natural Earth

Chapter One

OLYMPIC NATIONAL PARK

Olympic National Park in Washington State is a land of towering mountains, old-growth rain forest, rugged seacoasts, and exceptional diversity. The tallest coastal mountains in the coterminous states rise to almost eight thousand feet, just a couple dozen miles from the Pacific Ocean, capturing bucketloads of its moisture in the form of rain and snow. Some of that snow transforms into glaciers—184 of them at last count, a number bested (outside of Alaska) only by another Washington national park, North Cascades. Down from the mountain peaks, the rain nourishes temperate rain forest, forests so lush that seeds struggle to find bare soil upon which to germinate. Water seems to be everywhere, from rivers roaring with a foot of winter rain to rivulets gurgling quietly under fallen branches, and even summer brings just a fleeting experience with aridity. The streams and rivers bringing the moisture back to the ocean sustain some of the country's strongest salmon runs (again, outside of Alaska); one of them, the Elwha, may soon see its historic runs of all five types of Pacific salmon return, thanks to the recent removal of two dams that had impeded them for decades (see plate 1). Overlooking the salmons' maritime home is the last of Olympic's magnificent landscapes, seventy miles of rugged, undeveloped coastline. Within these landscapes is a wide variety of plant habitats, from wet to dry, submerged to frozen, and tall to diminutive—together sheltering over 1,450 species of plants. The Olympic landscape is, indeed, a place of diversity, beauty, and boundless wonder.

And roads just scratch the surface of this million-acre preserve, leading one to the edge of its sprawling wilderness. No roads cross the park's mountainous interior, so most visitors skim the edges, driving up to Hurricane Ridge for its mountain panorama, walking the nature trails in the Hoh Rain Forest, and getting their feet wet at one of the park's Pacific coast beaches. Few venture beyond the roads and into the park's wilderness, where the richest rewards are found. In 2013, I did just that, meeting my friends Jeff Pappas and Brenna Lissoway there for a five-day, forty-mile hike that brought us within shouting distance of the park's largest glacier. It was a personal, sometimes arduous introduction to several of the park's most celebrated resources—and, by hike's end, a frightening indication that my body was under attack from something powerful and insidious.

Our adventure began with a full day of hiking through rain forest, following the Hoh River upstream from the popular nature trails. It's hard to imagine a place in the temperate zones of the planet more brimming with life than Olympic's rain forests, found in its western valleys. Open to the full strength of Pacific Ocean moisture, the Hoh gets 140 to 167 inches (that's twelve to fourteen *feet*) of rain every year, making it the wettest place in continental America. The Hoh also enjoys the moderating influence of ocean proximity on its climate, rarely seeing freezing temperatures or summertime highs above 80 degrees Fahrenheit. This wet and temperate climate, along with fertile, alluvial soils, makes for an explosion of greenery, in size, diversity, and competition. Sitka spruce, western hemlock, and Douglas fir dominate the canopy, attaining heights of two hundred feet or more, circumferences of thirty to sixty feet, and ages of hundreds of years. These trees supplement summertime rainfall with the moisture from fog condensing on their needles, getting up to 30 inches of additional moisture this way—and giving the lie to summers that are surprisingly dry, for a rain forest. Below the heights, big-leaf maple adds deciduous beauty, with 130 species of ferns and other epiphytes festooning every branch, some of them hanging down in living draperies. Some of the epiphytes fix nitrogen (convert atmospheric nitrogen into a form plants can use), so the maple puts roots into them—aerial roots identical to those underground. No open ground (other than the trail) is present, with over a thousand species of mosses, lichens, and liverworts competing for every square inch of living space. So coveted is that space that fallen logs become tree nurseries, lines of saplings whose roots eventually belie the once-nurturing log, now rotted away. In some areas of forest, every last tree got its start on a log.[1]

Finding the moist, vibrant conditions to their liking is an assortment of wildlife, including thirteen different amphibians, one of which, the Olympic torrent salamander, is endemic to the park, found nowhere else in the world. Entomologists surveying a similar rain forest on nearby Vancouver Island in British Columbia found fifteen thousand species of insect, including five hundred new to science—in just the canopy alone. At the other end of the size scale are Olympic's Roosevelt elk, five thousand of them park-wide. Although we would not see any on this hike, we certainly saw evidence of them in the open forest and in the dominance of Sitka spruce. Elk prefer the needles of western hemlock and Douglas fir in winter (along with other more tender plants) to spruce needles, which are hard on the elk tongue, giving the advantage to the spiny spruce. Not only are these forests bursting with life, but they are also overflowing with age—and the living things that depend upon centuries-old conditions. Many of Olympic's rain forests have not seen fire, the most common disturbance in most western forests, since 1308, when fires burned half the forests on the Olympic Peninsula. Other fires burned in the mid-1400s, up to 1538, but there are many Douglas firs approaching their seven hundredth birthday (Douglas fir is the first tree in these forests to reestablish after a fire). Also celebrating antiquity (proverbially) will be a lichen that is only found on trees over four hundred years old, and forty different animals that depend on old-growth forests to such an extent that their needs would not be met otherwise. Olympic's rain forests are indeed a profusion of life and age; hiking in them is like walking through a cathedral, conversing in hushed voices and feeling like anything faster than a reverential stroll is an impertinence.[2]

After ten miles of hiking through the rain forest cathedral, we reached the Olympus Guard Station, which was our signal to set up camp in the nearby campsite. We pitched our tents on the banks of the Hoh River, a premier spot with views of the High Divide, a lofty ridge we would climb three days hence. We passed a pleasant evening watching to see if a persistent cloud would separate itself from the highest crown of the ridge; it did not, so the High Divide's full identity would remain a secret to us for another day. The gurgling river, here the size of a large creek, was as much a presence as the divide; its clear water was bounded by banks of gravel and alder saplings, the tree of rivers throughout the West Coast. With the route promising a 3,400-foot climb the next day, we turned in, the better to get a full night's rest.

Except for me—waking up refreshed would be unlikely, since I had been unable to sleep deeply for a week or so. I would sleep for an hour or two, then wake up, feeling tense and on edge. That would be the pattern for the entire night: sleep for no more than two hours, wake up, turn over, wait for sleep to overcome my tension. I was beginning to suspect the antidepressant medication my neurologist had prescribed; actually, I was beginning to suspect the doctor himself, for several reasons. He'd been dismissive of my concerns from the start, ignoring the two-page patient history I'd prepared and telling me after a cursory physical exam that I didn't have either MS (multiple sclerosis) or ALS. Instead, he interrupted me—as I went through the patient history anyway—to tell me I just had general anxiety disorder and then put me on an antidepressant. A month later, when I called his office complaining that the medication was not doing anything to alleviate my symptoms, he had his aide tell me to come in. But then, when I did see the doctor, he asked me why I was there, as though I were wasting his time. After telling him he told me to come in, he put me on Cymbalta, another antidepressant. It didn't seem to be doing anything to remedy the symptoms either, though it was too early to tell for certain. Much like our ignoring the signs that the climate is changing—with some (perhaps many), willfully ignoring the evidence—the neurologist chose again to ignore the indications that my body was actually in the early stages of something big, perhaps life changing.

My symptoms had begun six months earlier. I had bought a house near Yosemite, where I worked as a project manager overseeing the development of a major management plan for the Tuolumne River. Work was stressful enough, but adding house buying (long known to be one of life's most draining events) to the mix made for an exceptionally stressful time. I was handling it okay, noticing only an odd feeling in the back of my throat. That went away once I moved in, but a couple of months later

I began noticing many other strange things. In addition to the symptoms I'd told Josh about four months earlier (slurred speech, twitching muscles, and crying easily), my left leg seemed slow, and both legs would spasm and shake involuntarily just before I got out of bed in the morning. Also, just before I left for this trip, I had resorted to using toenail clippers to trim my fingernails, my fingers becoming too weak to drive the shorter-handled fingernail clippers through their intended targets. I had put weakening fingers on the patient history as a possible symptom, noticing that my handwriting seemed a bit messier than normal, but I wasn't sure about it. Now it was certain—and I would soon be telling that to a different neurologist. Fed up with the insensitive neurologist, and listening to my family's urging me to get a second opinion, I had scheduled an appointment with the neurology department of Stanford University Hospital. It would be shortly after this hike—and weakening fingers wouldn't be the only update to the patient history I would be making.

Back in Olympic, we had a two-day side trip to make. Once we climbed the 3,400 feet that had sent us to bed early on the Hoh, we would be in Glacier Meadows, the closest a person could get by trail to Mount Olympus, the highest peak in the park at 7,980 feet above sea level. When we had decided to go to Olympic and do an extended hike, I had volunteered to choose a route. We wanted a good cross section of the park, and I wanted to get as close as possible to the mountain, having failed to see it on a previous hike there due to cloud cover (a not-infrequent occurrence in Olympic). That meant we'd be hiking the Hoh River Trail, but it comes to an end just past Glacier Meadows. To give ourselves some variety on the way out (instead of retracing our steps all seventeen miles along the Hoh River Valley), we would hike up and over the High Divide, coming out the Sol Duc Valley to the north. This would make for a rugged five-day hike, but we were all fit and the route had plenty of rewards for our exertions.

As we climbed along the upper Hoh, the forest gradually changed, with Sitka spruce dropping out and silver fir and western hemlock taking its place. We crossed the river on a bridge high above (so high we looked down on mature trees), and then began ascending Glacier Creek, a tributary with a fitting name, as we would soon see. We stopped at Elk Lake, a pond surrounded by old-growth forest that made for a pleasant lunch stop and refreshing dip. Resuming the climb, we soon arrived at a steep ravine at which the trail ended abruptly, reappearing on the other side. The ravine is in an area of fragmented rock that is inherently unstable, making it difficult to build a trail that lasts more than a year or two. To facilitate crossing it, the park service installed a cable and wood ladder, laid onto the sloping side of the ravine to the bottom, and another such ladder on the other side. Once across the ravine, we soon broke out of the trees and had a view that was a down payment of sorts for what we were soon to see, up close and personal. Across the valley, no more than three miles away, was White Glacier (see plate 2). This river of ice begins in a large snowfield on the western shoulder of the mountain and sends a seemingly delicate snout, for so big a snowfield, down toward Glacier Creek, stopping a mile and a half from it. Audible from the trail, though, and linking the glacier to Glacier Creek, was the glacier's meltwater stream, tumbling down the steep side canyon the glacier had carved out in cooler times. The sight whetted our appetites for what lay ahead.

Soon we arrived at Glacier Meadows and set up camp. I was disappointed by the lack of a view from the meadows; I had envisioned an expansive meadow framing a massive, snowy Mount Olympus, but instead there was a small, sloping meadow, closely bounded by forest. No mountain was visible at all; we could almost have been in any state that has conifers. However, I knew there was a viewpoint a little farther up, so with a couple of hours on hand before suppertime, I asked the other two if they wanted to take the time to go to it now. They were more interested in taking a nap, but encouraged me to go there and report back to them for the next morning. I took off, and quickly found the trail was more of a climb than a flat stroll through the park. It soon emerged from the forest and began climbing toward a saddle, or a low point in the ridge whose side I was climbing. Late summer wildflowers were sprinkled among the grasses, but the vegetation soon gave way to rocks. A few drifts of snow were still present, their meltwater starting a rivulet trickling among the boulders. There was still no sign of the mountain, though it was clear I wasn't in Kansas anymore.

The afternoon was warm, almost hot, so when I learned from a hiker descending from the overlook that a glacier was prominently visible from it, I quickened my step. A short while later, I was there, looking down on Blue Glacier and up to the three summits of Mount Olympus (see plate 3). Like its White Glacier cousin, Blue Glacier gets its start in a huge snowfield by the three summits, but Blue stands out for being every bit the size one would expect from such an expanse of snow and ice. From the broad snowfield, the nascent glacier breaks into three or four parallel icefalls that then all cascade down hundreds of feet alongside each other, complete with crevasses in the ice and arêtes (knife-edge ridges, carved by parallel glaciers) between the icefalls. Below them, the frozen streams gathered together as one mighty river of ice, flowing down—and filling, wall to wall—the valley below the overlook, turning left as gracefully as the icefalls spilled chaotically. The overlook itself is actually on a lateral moraine, a ridge of boulders, sand, and gravel deposited alongside the melting glacier. All told, Blue Glacier is about two and a half miles long, descends about three thousand feet, and is the longest and one of the largest glaciers in the park.

Glaciers form anywhere that receives more snow than can melt in the summertime. Blue Glacier receives more precipitation than any other glacier in the forty-eight contiguous states (240 inches annually, compared to 30–40 inches for most states east of the Great Plains), largely because it's so close to the Pacific Ocean and because Mount Olympus is tall enough that the moisture in the clouds freezes as they move inland—and up—from the lowlands like the Hoh Rain Forest. Some fifty to a hundred feet of snow falls on the mountain every year, producing so much ice that the four largest glaciers in the park are all on Mount Olympus (Blue, White, Hoh, and Humes; the latter two we would not see); there are nine glaciers in all, containing 80 percent of the park's ice.[3] If a glacier receives less snow than melts annually, though, it shrinks or retreats. All of Olympic's glaciers have been retreating since the late 1800s, with Blue Glacier's terminus receding 0.7 miles since then. The glacier is also getting thinner, losing 178 feet of depth in its terminus, and 32–48 feet in its snowier upper portion between 1987 and 2009 (which explains why the lateral moraine and overlook are above the glacier: it was once as high or higher than them). Park-wide, the surface area of glaciers shrank by 34 percent between the 1970s and 2009, with the actual number of glaciers dropping from 266 to 184. Glaciers in the drier northeastern part of the mountain range (such as Lillian Glacier, which is gone) or on south-facing aspects (such as Anderson Glacier, which has lost 77 percent of its area) have been the hardest hit. Finally, the rate of melting is increasing; the rate since 1980 is double that of the previous eight decades, and the rate since 2009, quadruple that (with another 36 glaciers melted out since 2009).[4] Since 1986, Blue Glacier has been retreating 50 feet per year, and precipitation at its terminus (which is about 4,500 feet) now falls more as rain than snow.[5] In sum, glaciers may appear massive and unchanging, but they are sensitive barometers of changing climates.[6] Moreover, their retreat is the most incontrovertible evidence that our climate is changing—worldwide, for Blue Glacier isn't the only one retreating. Most other glaciers in the West and, indeed, around the world are also retreating, clear evidence that the global climate is warming.

Global warming begins with carbon dioxide and methane, two relatively rare gasses in Earth's atmosphere, known collectively as greenhouse gasses for their effects on Earth's climate (there are other greenhouse gasses, but these two are the most important). They allow the sun's visible light in, but trap most of the infrared radiation that Earth radiates back to space. This is the "greenhouse effect," whereby some of the sun's warmth is retained on Earth. Greenhouse gasses are powerful drivers of the climate we enjoy; without them, life on the planet would be much more difficult, because most of the sun's warmth would be lost, especially at night. Problems arise, however, when we add more greenhouse gasses to the atmosphere, largely by burning fossil fuels like coal, oil, gasoline, and natural gas. With more heat-trapping gasses in the atmosphere, the planet warms a little more, and this process continues to accelerate the more carbon dioxide and methane we contribute to the atmosphere. Since the Industrial Revolution, the carbon dioxide concentration in the atmosphere has been increasing (from 280 parts per million to 409 parts per million in 2018), and the global temperature has followed in lockstep: globally, we're now 1.8 degrees Fahrenheit warmer than we were in 1880.[7]

Before moving to a discussion of climate change in Olympic, let's define a few terms. First, because this is a book about national parks in the United States, and will primarily be read by Americans, the temperatures will all be given in degrees Fahrenheit. Second, the book is about changes in the climate, not weather in the parks. If the two were the same, one might conclude that International Falls, Minnesota, has a warm climate because it recorded the nation's *high* temperature one July day in the early 2000s (the date escapes me, but it's not important): a torrid 99 degrees. It's common knowledge, though, that the city is one of the coldest places in the coterminous states, regularly setting the nation's low on any given winter night. Weather is what we experience on short time frames, from hours to days, weeks, and months, while climate is the precipitation and temperatures that characterize a place over years, decades, centuries, and millennia. Finally, the terms "climate change" and "global warming" are essentially synonyms today. Climate change was originally used in discussions of how humans were altering the climate, because some areas were not warming or even cooling, such as the American South. By the end of this century, though, there will be few—if any—such places. Consequently, global warming is more descriptive of the anthropogenic heating that is expected to occur everywhere in the coming decades. Even this term seems to understate the changes to come; I think "global heating" or even "global scorching" would be more fitting, but we'll stick with the common term, synonymous with "climate change."

Now let's go to the changes that Olympic has already observed. Given its proximity to the Pacific Ocean, which somewhat buffers the park from the warming that more inland parks have seen, Olympic has only warmed by 1.2 to 1.8 degrees since 1900, the least of the five parks in this book. Still, the frost-free growing season is getting longer, spring is beginning earlier, and the coldest night of the year is getting warmer.[8] Temperatures in the higher elevations have risen more; those at Blue Glacier, for example, are 5 degrees warmer since 1948.[9] These changes may seem trivial, but they have been enough to melt off some eighty glaciers, along with causing a host of other changes. The water content of the winter snowpack, known as its snow water equivalent, or SWE (as measured on April 1), is now 15 percent less than it was when measurements began in 1968, with some lower-elevation sites declining markedly more.[10] Spring runoff is also happening earlier and soil moisture is being recharged earlier, both the result of advancing springs.[11] Precipitation patterns are changing in other ways, too, with spring precipitation increasing while summer and fall rainfall are declining.[12] Storms are becoming more intense (because heat is energy and warm air can hold more moisture than cool air), producing more, and higher, floods.[13] The floods have led to river-channel widening downstream on at least four rivers draining the park, including the Hoh.[14] Overall, though, annual streamflows are declining, and dry years are becoming drier.[15] Perhaps as a result of this drying trend, background mortality rates of most western forests have increased since 1955, including those in Olympic.[16] Finally, one group of researchers has demonstrated that human activity (burning fossil fuels) is the primary driver of many of these changes, while another group showed that the same activity made a recent drought year (2015) up to 33 percent worse than it would have been if natural factors alone were at play.[17] The human signature in causing these changes is clear.

Already, some plants and animals may be responding to the changing Olympic climate. Trees have been moving into subalpine meadows and alpine tundra for decades, indicating that the temperatures there have warmed enough to make the climate tolerable.[18] Songbirds, too, may be benefiting from a changing climate; a recent population survey of thirty-three songbirds in Olympic found all but one either stable or increasing in density. Reduced snowfall is the likely reason for the increasing densities, a finding of other songbird-monitoring efforts nationwide. The one species that declined, the Olive-sided Flycatcher, did not show a corresponding increase at higher elevations in Olympic, though it did in Mount Rainier, where there is more high-elevation habitat.[19] The counterintuitive finding of increasing songbird density, while welcome for now, is unlikely to persist as our emissions continue to warm the planet. Likewise, the warmer climate facilitating the movement of forests upslope will probably not be so friendly to the forests left behind at lower elevations. In other words, we should not be sanguine about the unexpected results, for they are just a hint of the warming and environmental disruptions yet to come.

Back at the Blue Glacier overlook, I thought about the prognosis for Blue Glacier: more, and probably increasing, melting and retreat, perhaps culminating in its disappearance. I kept the thought in the back of my mind as I looked, awestruck, at Blue Glacier and Mount Olympus that warm afternoon, and again as I shared the view with my two backpacking partners the next morning. The three of us actually got a predawn start so we could watch the sun rise over the mountain and its icy progeny. We had the overlook to ourselves, and arrived just after the sun had begun to highlight the mountain's arêtes, upper glacier, and conical summits. The first rays were orange, the air was calm and quiet—a profound stillness, such as even the wilderness rarely provides—and the temperature was cool but not cold. The three summits were a deep umber, with a number of lesser, pointy false summits also catching the sun's first rays, the lot of them looking much like a frosty castle of the abominable snowman. Below the yeti's minarets sprawled the snowfield, an unbroken blanket of peach-colored snow serenading the morning sun. The snowfield's smoothness was an indication that the summer sun had not yet melted last winter's snow, despite the fact that the summertime melt was more than half over. Rupturing the snowfield's continuity were the icefalls, their crevasses and the reunified glacier below appearing a neutral gray in the still shadows, the natural complement to the shades of orange above. The icefalls and reunified glacier showed a more typical mix of old and new snow, with the new, clean snow hanging on in sheltered, shady areas, while the ice in more open places was dirtier, with the accumulation of rocks and other debris throughout. The mountain and glacier were an evolving show of colors, so we absorbed the show and its energy through coffee and breakfast, surely one of the more memorable meals of our lives. It felt like a scene out of place, for until arriving at this overlook yesterday, I personally had not associated Olympic National Park with glaciers, let alone glaciers big enough to look like rivers of ice. I knew the park had tall mountains along with its rain forests and rugged coastline, but the scene was a revelation, an obvious statement that this park preserves many resources of significance.

Glaciers are not just barometers (thermometers!) of a warming planet—they're also treasure troves of information about past environments on Earth. Perhaps the best example of this is their record of past greenhouse gas concentrations, which have varied over time, along with the warming Earth has seen, as revealed in ice cores from Greenland and Antarctica. As snow falls and accumulates over time, it traps bubbles of air and preserves them as it is compressed into ice. By drilling out a vertical column of ice, cutting it into thin cross sections corresponding to years (determined via a number of methods), and then melting the layers and trapping the air liberated from the bubbles, scientists can determine past greenhouse gas concentrations. Six hundred fifty thousand years of ice are preserved in the two ice-bound places, an exceptional window into past climates (not to mention a great form of job insurance for the people who analyzed all those cross sections!). The results were sobering though: today's concentration of greenhouse gasses is the *highest* in the last 650,000 years. The ice cores also tell us that the contemporary rate of change for the two gasses is the *fastest* in that same time period. Roughly ten times faster than the warm-up that melted the huge continental glaciers of the last ice age, the current rate of warming will outstrip the ability of many plants and animals to follow their preferred habitat northward and/or upward in elevation. Finally, the ice cores also tell us how warm or cool the planet was over time, based on the relative concentrations of two different oxygen isotopes. Like all elements, oxygen molecules have electrons circling around a core of protons and neutrons. In the case of oxygen, most molecules have eight protons and eight neutrons, but a small percentage of them have ten neutrons, making them heavier than the other molecules. The relative abundance of the heavier molecules correlates with the relative warmth of Earth, with higher concentrations indicating a warmer planet and lower concentrations indicating a cooler time period, such as an ice age. By correlating this timeline of relative planetary warmth to the record of carbon dioxide levels in the same time period, scientists see a one-to-one match: when greenhouse gas concentrations were high, the planet was warm, and vice versa. The ice cores, then, not only validate the premise of the greenhouse effect, but they also illustrate that the human signature on contemporary greenhouse gas concentrations is remarkably clear.[20]

Scientists communicate such findings to each other through articles in various technical journals, books, and reports, including government documents. All such literature is peer reviewed, which puts the research to the scrutiny of the scientist's professional peers (usually two or three) before the publisher accepts the manuscript. The reviewers, chosen by the editor for their expertise in the relevant field(s), look to see that the methods are appropriate for the question being studied, that they are repeatable, that the results support the author's conclusions, and that the research makes an original contribution to the body of knowledge in the applicable field. Once the reviewers provide their comments to the editor, the author is expected to address them or rebut them with justification, or go back to the drawing board. As contentious as any climate science is today, such research is subjected to an even higher level of critique, thereby ensuring that the published research is accurate and relevant. Almost all of the climate science research upon which this book is based has been peer reviewed, along with this book itself, ensuring that the information presented here is as accurate as possible.

The peer-review process is intended to prevent erroneous, flawed, and junk science from being published. For example, once my book *Yellowstone and the Snowmobile* was published in 2009, a few editors asked me to review articles about other facets of motorized winter recreation and its management. One such manuscript was little more than a xenophobic screed against federal attempts to control unbridled snowmobile use on federal land, so it was obvious I should recommend against its publication. Like any system to ensure quality, peer review can sometimes allow problematic research to find its way into print. Here again, personal experience illustrates how the system corrects itself. In the 1990s, Yellowstone National Park's managers faced ongoing criticism over their management of ungulates (hoofed animals), especially elk. One critic, a Utah State University professor, was particularly persistent, publishing several articles asserting that prehistoric elk numbers were much lower than contemporary ones, owing to humans. I followed his paper trail, finding most of his claims baseless or exaggerated. Once I published my critique of his claims, his criticism no longer held much water.[21] In these and other ways, the peer-review process assures that all climate science is accurate as possible and incorporates the information and research techniques of the era.

Based on many lines of such peer-reviewed evidence (like the ice core information, and other evidence to be discussed later in this book), scientists are virtually unanimous in their assessment that human activity is the primary driver of the increase in carbon dioxide levels. Since 2004, a number of researchers have gathered as many peer-reviewed papers as they could find reporting research into climate change and global warming, and assessed the percentage of authors accepting humans as the primary driver of global warming, having no opinion on it, or rejecting the human role in global warming. In every case but one, the percentage of scientists accepting that humans are at least partly responsible for global warming was 100 percent, or extremely close to it. Even the outlier, whose methodology overestimated the number of skeptics, still found that 97.1 percent of scientists were in agreement about the human role in causing climate change. The most comprehensive of these papers is a 2017 inquiry that looked at 54,195 papers published between 1991 and 2015, finding just 31 papers rejecting the human role in global warming, representing 0.06 percent of the total. Tellingly, the 31 rejecting papers have only rarely been cited by other scientists—even other skeptics—and several have never been cited, indicating other scientists distrust them. While many of the 54,195 papers expressed no opinion on the subject (many of them reported on methodology or other topics relevant to the debate), it's hard to find a subject on which scientists are speaking with one voice—99.94 percent!—more uniformly.[22] Indeed, the case for humanity's burning of fossil fuels as the primary vector of global warming has been thoroughly vetted, reviewed, and accepted in the scientific community. For that reason, no national or international scientific organization rejects the premise that humans are the main cause of global warming.[23]

Meanwhile, at our wilderness breakfast table, we savored the silence and scenery till the sun reached us. That afternoon, we retraced our steps to the guard station and campsite on the Hoh River. I took a short side trip on the way down to the glacier's terminus, its snout. Sloping downward as the ice approached its end, the glacier gave way to rock, both bedrock and loose rocks. The bedrock was smooth in places, polished by the ice no longer there, and the glacial debris varied in size from sand and gravel to cobbles and two-foot-diameter boulders, all plucked from Mount Olympus and brought here by the glacier. No meltwater stream was present there, meaning it's subterranean, flowing through the glacial debris and emerging a short distance downhill. That stream, along with the other Hoh River tributaries draining the Mount Olympus glaciers, gets a sizable portion of its summertime water from melting ice, especially in August and September, when up to a third of the Hoh River's water is glacial meltwater. It's easy to see how the glacial melt-off could affect streamflows and the aquatic life dependent on them.[24]

Water isn't the only thing that the Hoh and its tributaries receive from the Mount Olympus glaciers; they also pick up sediment, so much that when the river's downhill gradient lessens near our guard station campsite, the river begins to braid. When a river carrying sediment reaches flatter terrain, the river no longer has the kinetic energy to continue transporting its larger sediments. The sediments settle out, which raises the riverbed a little. The process continues until the riverbed is as high as the surrounding land, at which point the river moves to one side or the other, whichever is lower. In so doing, the river will take out the trees growing on the lower riverbank, while other trees will germinate on the newly abandoned streambed. The Hoh braids more than any other river in Olympic (because 65 percent of the park's glaciers are in its drainage), giving it a unique profile in the Olympic Mountains, one that will probably disappear when the glaciers melt out. At that point, the river will transform into the more typical, mature form that characterizes most of Olympic's rivers, with stable banks and mature vegetation. This will result in a loss of riverine diversity, along with the loss of any aquatic insects and microflora and -fauna that prefer the open and warmer conditions found in braided streams.[25] In these ways and more, glaciers and their meltwater (natural and human caused) can have ripple effects far downstream.

As noted previously—and on a far bigger scale—glaciers around the world are melting, including those in Greenland and Antarctica, contributing to sea-level rise. The oceans are warming as well, further contributing to rising ocean levels (because water expands as it warms). So far, though, the Olympic coastal strip has seen the opposite occur, because the coastline of both Oregon and Washington is rising as the Juan de Fuca Plate is forced under the two states. Olympic's stretch of magnificent, rugged coastline is one of just two places on the Pacific coast where a person can take a multiday backpacking trip, walking on the beach, and not see a human structure (the other is California's Lost Coast). I was fortunate enough to make such a trip in 1998, immersing myself in the marine air for three days of glorious solitude. But for a few short stretches, my trail was the beach, and my only companions were the deer I saw wading into the surf at low tide to nibble on kelp and the plants growing in tide pools—oh, and the banana slug I inadvertently rolled up with my tent when I broke camp before dawn on the last day (to get around a headland before high tide). Squished slugs aside (my tent is still stained), the memories from those three days are images of wildness: the tempestuous Pacific, the sea stacks, the tide pools, the marine deer, all without a road, motors, lights, or vehicle noise.

It won't be long, though, before sea-level rise begins to outstrip the rising land, removing a bit of the wild nature I'd found there. By 2030, scientists estimate the ocean will be two inches higher, and by 2050, six inches, differences that are mostly within the present-day variations in sea and tide level. But by 2100, the projected sea level, assuming business continues as usual, will be twenty-four inches higher in Oregon and Washington. That would be enough to cause significant erosion of the shoreline and, in flatter areas, flooding, especially during king tides (the highest high tides). Rising seas will affect dozens of cities and hundreds of millions of people worldwide; for example, both the San Francisco and Oakland airports will begin to see their runways go under when sea-level rise gets to sixteen inches. Without a subducting continental plate, California's coast is expected to see a three-foot rise in sea level by 2100—a foot by 2030—so the airports may be elevating their runways within our lifetimes.[26] Elevating runways will probably be the least of our worries when the oceans really rise, given the numbers of people who will likely be displaced by rising waters, with as much as eight feet of more global sea-level rise projected by 2100 (on top of eight inches we've already raised global sea level, and again assuming continued business as usual).[27] Compared with the widespread impacts on humans worldwide, rising-ocean impacts on Olympic's coastal strip may seem trivial; even within the park system, there are parks like Everglades National Park and Cape Hatteras National Seashore facing complete inundation due to unchecked sea-level rise. That does not diminish the effect on one of Olympic's treasured resources though: a two-foot rise in the Pacific Ocean would remove that which makes the coastal strip worthy of protection in the first place—its wildness.

After some downtime and a lovely evening on the Hoh River, it was time to tackle the second major ascent of the journey, a four-thousand-foot climb to the High Divide, the lofty plateau overlooking the Hoh River Valley. Not long after beginning the climb, we entered a burned area. In the summer of 1978, lightning started a fire on this steep, south-facing slope that just smoldered for two weeks until some dry east winds arose, increasing fire activity and enabling it to get into the canopy. For two days the wind blew, and for two days the fire grew, raging across 1,050 acres by the time firefighters contained it on August 14. It then went back to smoldering, not going out fully until that December.[28] By the time we passed through it thirty-five years later, the burned area was thickly vegetated, though I don't recall many trees. They were probably there, but likely not that tall, for I do remember abundant sunshine and wanting to get out of it. Seedling establishment has probably been slow, hampered by a lack of seed sources and competition from faster-seeding shrubs. Soon enough, though, we climbed out of it and were taking our packs off for a welcome break at Hoh Lake, a scenic tarn marking the beginning of the High Divide. From there it was just another thousand feet of climbing to the backbone of the divide.

Fire is a natural part of most western forests, including the ones we hiked through on that hike—even the temperate rain forests (if infrequently). It's so much a part of them that the trees tend to adopt one of three strategies to cope with it. The first is to survive fire, usually by producing bark thick enough to insulate the cambium (their growing tissue) from the heat of the flames. Trees that see periodic, low-intensity fires, such as fires every three to twenty years, tend to show this adaptation. Giant sequoia trees take this strategy to its limit, producing bark that is one or two feet thick! Other trees like Douglas fir and ponderosa pine produce two-to-four-inch-thick bark (a more common thickness). The second strategy is akin to life insurance, and is typically found in forests that see stand-replacing fires every one to three hundred years: serotinous cones. These are cones that don't open to release their seeds until the tree is burned. Lodgepole pine is a good example of this strategy, with crown fires that kill the trees but open the serotinous cones, starting a new forest. The third strategy—to anthropomorphize the tree—is to hope fire doesn't hit, and that if it does, it will be patchy enough to skip over you so you can reseed the area. This strategy tends to be found in areas that burn so infrequently that producing either of the other two adaptations is not a productive use of the tree's resources. Western hemlock, the primary tree here before the fire, employs this strategy, as does Sitka spruce. There are other adaptations to fire, such as sprouting from the stump of the burned tree, but these are the three main strategies that trees in the western coniferous forests tend to display.

Fire-return intervals in Olympic are thought to be as much as nine hundred years in Sitka spruce forests, and can be very powerful and destructive when they do occur.[29] A look at the Tillamook Burn in Oregon, which occurred in a similar forest, provides a glimpse into what a fire in Olympic's rain forests could look like. In August of 1933, logging operations in Gales Creek Canyon fifty miles west of Portland ignited a fire that burned about forty thousand acres over the next ten days. Then, on August 24, a strong east wind arose, pushing the fire across an additional two hundred thousand acres in just twenty hours, one of the fastest rates of wildfire spread in the twentieth century. Flames exceeded a thousand feet in height, and heat from the fire was so intense that smoke rose into the stratosphere, forming a thunderhead approximately 40,000 feet tall, visible from as far away as neighboring Washington. What's more, the area burned every six years after that until the 1950s, in 1939, 1945, and finally, 1951. All three of the subsequent fires were also started by logging operations, and reburned much of the same area. Dead lumber may have stoked the reburns, but the brush and grasses that typically revegetate burned areas before trees get established probably provided most of the fuel for the subsequent fires. With the reburns killing the few seedlings that had germinated after the original fire, seedling trees were scarce in the later years. After the last fire, the state of Oregon took a more active role in reseeding and revegetating the area, to the point that today the area is a state forest, though signs of the fiery past remain if you know where to look.[30]

Olympic's rain forests are wetter than the Tillamook area's, but the similarities are otherwise striking. Some of the trees are the same, the climate is similar (with cool, wet winters and warm, dry summers), and east winds can be desiccating. Could such a fire or fires happen in the Hoh drainage or Olympic's other lush forests, and how will global warming influence the likelihood and severity of such fires? To answer questions like this, scientists turn to climate models, which are computer programs that climatologists devise to predict the future climate under different scenarios. The models employ the various quantifiable relationships between the various components of the climate (think of sophisticated mathematical formulas like the laws of physics) to create Earth's climate in digital form. With climate consisting of the atmosphere itself and the various ways that it interacts with the land, oceans, ice caps, sunlight, and human activities, and with climate evolving over more than a century, beefy computers are generally necessary to run the models. Once a scientist has a basic model set up, they fine-tune it by going back in time and adjusting the formulas to make it predict the past climate accurately. For example, the 1992 eruption of Philippine volcano Mount Pinatubo affected the world climate for three or four years, giving scientists an opportunity to tweak the model in response to such events. Once the model is successfully "hind-casting," as it's known, the model is assumed to be accurate and the researcher can turn to future projections.[31]

Human activities, one of many factors in climate modeling, come to play pivotal roles in modeling the future, especially when we're predicting the climate several generations from now. Because there are and will be myriad choices we'll make that affect the climate between now and the end of the century (the typical end point of the projections), climatologists model a suite of different scenarios. However, to simplify the ensuing discussion, they usually just present two scenarios that present the extremes of possible outcomes, a business-as-usual option where we do little to address global warming, and a concern-for-the-future option where we proactively take steps to reduce our carbon footprint (the sum total of the per capita greenhouse gasses we emit). Projections are often broken down by seasons of the year, which provides a more complete picture of likely effects. Winter projections reveal probable effects of climate warming on snowfall and phase of precipitation (snow versus rain), while springtime changes affect snowmelt and plant emergence. Summertime changes affect forest drying and the likelihood of wildfire, while fall extends the effects of both summer and winter changes. This characterization is generalized; for example, changes in winter snowpack have a direct effect on summertime fire risk. Also, models typically have strengths and weaknesses, given the assumptions that go into them, the purpose in creating them, and the complexity of the actual climate. One model may do a better job predicting summertime extreme high temperatures while another one is better with winter snowfall. For this reason, researchers typically use an ensemble of twenty or more models, all run with the same assumptions about human activity, in projecting the future climate.[32]

Climate modeling has been growing in sophistication and specificity, becoming more and more accurate and allowing us to predict the future climate in specific locations. There are several websites that allow one to derive the projected climate for a location of the user's choosing, using the same models that research scientists use. One of them, the Northwest Climate Toolbox, is sponsored by a group of government agencies and climate change research consortia, and includes a variety of climate monitoring applications and climate change projection tools. Using the Hoh Rain Forest and Mount Olympus as our two locations (because they best represent the focal points of our discussion, rain forests and glaciers), we see that both locations will warm by about 10 degrees by the year 2100. This warming is fairly consistent across the seasons and metrics (such as average annual temperature, average overnight wintertime low temperature, average summertime high temperature, etc.). However, two metrics deserve specific mention, and both occur under the business-as-usual scenario (in most cases, the concern for the future scenario limits the temperature increase to only about half of the other). The first metric of interest is on Mount Olympus, where the typical winter day and night will move from below freezing to above (average low temperatures going from 22 degrees in the 1950s to 33 degrees in 2100, and high temperatures moving from 32 degrees to 41 degrees in the same time period).[33] Snow will still fall in that climate, with some research indicating that, when snow is falling heavily, it can effectively drag the cloud's colder temperatures with it, overcoming warmer surface temperatures.[34] However, once the storm moves on, temperatures would revert to those of business as usual, so the melting would return.

The winter of 2014–2015 provided a glimpse into a future in which these projections become an uncomfortable reality. That winter brought temperatures 4 degrees above average, and the months of January to March (the heaviest snow months) were all 6–7 degrees warmer than average. The winter saw close-to-average precipitation (93 percent), but most of it was rain, and the pitiful snowpack that did accumulate melted off completely in the face of February's warmth. It never measured more than six inches of SWE, in a place that ordinarily sees forty inches of SWE by winter's end. The influence of temperature is crucial; just two winters earlier, temperatures averaging 0.8 degrees warmer than average brought 106 percent of average snowpack, with over fifty inches of SWE by April. Three degrees has a powerful influence in snow country, as we'll see time and again in the western landscape—and that's just a third of the projected increase by the end of the century. Three degrees also had an outsize impact on the Olympic landscape, with Hoh Lake remaining ice free most of the winter, dropping from 226 days frozen in a typical winter to just 81 days iced over amid the warmth of winter 2014–2015. More affected were the park's remnant glaciers, which lost about thirty feet of ice that sneak-peek winter, especially the lower ones.[35] The future of Olympic's glaciers appears bleak indeed.

The other metric is in the Hoh Rain Forest, where the summertime average high temperature projected by the Northwest Climate Toolbox will increase from a cool 69 degrees in the 1950s to almost 80 degrees by 2100. Such increased warmth will promote the growth of any fires that occur, assuming rainfall totals are near today's amounts.[36] Unfortunately, that assumption is probably not a safe one. Summers in Olympic are surprisingly dry, receiving less than a tenth of the year's precipitation (about 10 of a total 140 inches). Global warming will increase the disparity, with annual precipitation projected to grow by a foot while summer rainfall shrinks by almost 3 inches (to 7 of 153 inches).[37] More of the annual precipitation will fall as rain, leading to smaller snowpacks that melt out earlier in the year, and more precipitation will arrive in short-term, intense, flood-producing events.[38] Climate models don't handle precipitation projections as well as temperature, but even if the precipitation projections are wrong and summertime rainfall increases, the substantial increase in temperatures will most likely negate the added moisture. This is because the sun has more evaporative potential on warmer days, and because plants transpire more—earlier in the year—as temperature increases, soils dry out earlier in the growing season, taxing plants the more.[39] Indeed, one group of researchers specifically projected soil moisture levels throughout the West using ten climate models, finding an 81 percent reduction in the April 1 SWE in the Cascades by the 2080s, with a concomitant reduction in soil moisture levels.[40] Another team used twenty models to project that lower-elevation drainages in western Washington would essentially see their snowpack disappear by April 1, and that significant reductions (38–46 percent of average SWE) would be widespread by the 2040s.[41] Given these facts and the climate projections, researchers expect fires to increase significantly both in number and severity in Olympic. Also, drought-stressed trees have less resistance against beetles that burrow into them, eating the cambium (the tree's growing tissue) and usually killing the tree. Consequently, such beetles will likely increase in Olympic, furthering the climate for fire.[42]

Using these projections and a global vegetation model, another group of researchers projects that, by century's end, many of the forests in western Washington State would switch from cool conifers to warm mixed forests. Large fires, which have been rare in the region, will become more common, suggesting that fire will be the catalyst of dramatic ecological change. In so doing, fire will change from being a minor factor in forest ecology there to a major one.[43] With fires already increasing at a higher rate in the Northwest (though from a lower base) than in the northern Rockies, these changes have likely already begun. Tellingly, both the number of new, large fires and their duration are linked with early spring melt-off.[44] That causal relationship is probably true in Olympic as well, where the acreage burned annually has been growing since 2000, with the largest single fire occurring in 2015 and the most acreage burned the following year.[45] The Tillamook Burn may well give us a window into the future, one that is uncomfortably charred and austere. Given current trends, we should prepare ourselves for the day when the Tillamook Burn becomes the Olympic story.[46]

Back on our hike along the High Divide, we enjoyed an invigorating swim in Hoh Lake, followed by a relaxing lunch on its shore. Drying off in the glorious sunshine, I studied the vegetation on the opposite shore. There, the ground sloped steeply upward about three hundred feet, almost all of it covered with the vibrant green of vegetation trying to capture the sun's golden light while summer still dawdled. Only about a fourth of the slope directly across from us was forested, and all the mature trees grew in six or eight clusters, all of them areas of rocky ground. Outside of those forested islands, few mature trees grew, probably trimmed every few winters by avalanches. I pondered what made the rocky areas such good arboretums: Did they somehow shield the trees from snowslides, or offer a more tolerable soil-moisture regime for conifers? I suspect it's more the latter, with the rocky ground offering the well-drained soils that most conifers prefer. To the right of the steep slope, though, more typical, unbroken forest grew, from the ridgetop down a gentler slope to within thirty or forty feet of the shore. That forest didn't appear rocky, so why do trees find tolerant the presumably wetter soil there if that's what confines them to rocky islands on the steep slope? Avalanches don't shear trees off the gentler slope, but the rocky islands seem insufficient defense against avalanches for the trees growing on the steep slope. And why don't trees reach the shore at the bottom of the gentler slope? Maybe it's too wet there too, but with several miles still to go and a thousand feet to climb, we didn't have the time to investigate these mysteries further. Hoh Lake and the slopes above it remain wondrous, places of beauty with many layers of meaning and interrelationship to unlock. Whatever the exact answer may be, though, it's ultimately governed by climate; snowfall, rainfall, groundwater, and the temperature regime all play pivotal roles in determining what plants are found in a given location. The needs of the individual plant species and soil characteristics play important roles. In short, all species have their own performance limits due to their own traits that make them suited to certain kinds of soils and need for moisture. But climate is what delivers the most fundamental resource for any plant—water—which is the most crucial determinant of vegetation anywhere. Thus, as climate changes, so too will forest patterns.

Another hour of ascending and we were on the rim of the world: the top of the High Divide. The trail stayed on the ridge crest or just left of a line of scattered trees dotting it. To the north (and on our right) rose Mount Olympus, guarding a yeti's dreamland of sprawling snowfield and stone parapets. Between us and the mountain, and filling the view east and west, was the Hoh River Valley, four thousand feet below us. Except for the braiding on the Hoh, the green forest was unbroken, extending for ten miles up- and downstream, and for thousands of feet up the mountainsides, to within a few hundred feet of the mountaintops. Trees by the millions reached skyward, some on the valley floor pushing two hundred feet or more in height while their cousins at treeline strained to reach an order of magnitude less. Rimming the valley five or ten miles east of us were the peaks of the Bailey Range, six- and seven-thousand-foot mountains that turn the nascent river westward. Again, though, it's the forest that dominates the eastern and western views: not a dead tree anywhere in sight, be they killed from fire, beetle, or logging. It's becoming rare these days to see such a vast green forest in the West, with forests in the Rockies showing widespread beetle infestations and associated fires,[47] and with forests elsewhere in the Coast and Cascade Ranges of Washington, Oregon, and Northern California being heavily logged[48] (note that the Hoh Fire was out of view, but even if it had been in sight, it would have been a minor interruption of the green canopy). It was a view into a world that climate change has not altered much—yet—and a poignant reminder of the treasures at risk of being lost a la Tillamook.

Meanwhile, to the north (and on our left), the land fell gently away to Seven Lakes Basin, an open, thinly vegetated land of hummocks and sky ponds that seemingly gave us a warm welcome. Snow—thirty-two feet of it in a typical winter!—is almost as much an influence on the High Divide as it is on Mount Olympus, even though it had largely melted off by the time we hiked through. The divide is around five thousand feet in elevation—treeline in Olympic, the line above which conditions are too harsh for trees to grow. It's warm enough for trees, but so much snow falls that, in combination with the ridgetop's north-facing aspect, it takes until July or August to melt off in a typical year. That leaves too little growing season for trees to make it there, unless they happen to germinate in a location that melts out early, like the very crest of the ridge, which doesn't face north.[49] Additionally, the snow here is heavy and wet, so it crushes any trees that do try to make a go of it. In place of trees was a profusion of wildflowers, with the blues of lupines dominating and mimicking the sky above us. Other slopes were bare or nearly so, likely because they had just melted out—in mid-August. It was hard not to be enchanted with the High Divide, especially since it gave us some "rocks" to put our scotch on that evening (snow, from a remnant drift I had stuffed in a water bottle). We enjoyed a frosty toast as we made camp at Heart Lake, another scenic tarn (see plate 6).

Shaped somewhat like a valentine, the lake could have been the twin to Hoh Lake, with a similar size and similar green slopes above. If indeed they were twins, they must be fraternal, because the slopes across Heart Lake were mostly without trees, large or small. Just a few grew near the top of the slope, where the snow had blown away. To the right, though, a familiar pattern prevailed: trees in clusters, on rocky ground. This time, however, the landscape had a possible clue explaining the clustering: a remnant patch of snow, high in the bowl containing the lake and in an area not forested. Perhaps, then, the causal factor is that snowmelt drains away faster in the rocky areas, giving trees an advantage there that they don't have outside the rocky islands. In the forbidding treeline environment, trees can't handle many stresses, and the drier conditions on the islands give them just enough of an edge to survive. Once again, climate plays the most pivotal role in determining vegetation patterns, delivering so much snow and snowmelt that trees struggle to cope. The prodigious snowfall forces them off the landscape except where they can mediate the climate by growing where soils are well drained. Nature's mysteries are fascinating to the observant—along with its beauties. Adding to the subalpine beauty here were three shades of green covering the land. The conifers were a dark forest green (appropriately), while the sea between the islands were two shades of green, the basic green in any child's crayon box, and a bright, almost chartreuse green covering the flatter meadows. It was a transfixing delight, Heart Lake and its surroundings, the perfect closing of a glorious day, both intellectual and beautiful. However, as with my symptoms, the winter of 2014–2015 is an ominous sign of a troubled future, one that will affect all things influenced by the climate, including the vegetation patterns at the two exquisite lakes we passed on that memorable day.

In fact, I had another unsettling reminder that very evening that something was going wrong in my body. When I inflated my air mattress, I found that I could not blow into it without holding my nose to keep the air I was blowing from escaping nasally. Whatever part of my soft palate that closes the nose off when inflating something could no longer seal it effectively, even though I hadn't had this problem the night before. That was actually the lesser of two significant omens; the other was that my speech had become noticeably slurred by that point in the hike, all the time. Until then, the only time it had been slurred was when I was really tired, and normal speech had always come back as soon as I got a good night's rest. Not this time though. The insomnia was still a problem, and as arduous as the two climbs of the trip had been, I needed—but was not getting—good, restful sleep. Slurred speech was evidently here to stay, and I could only hope it would disappear once I got a good night's rest after we got to the car and slept in a comfortable bed.

Unfortunately, that would not be the case for me. Slurred speech stayed with me for the hike out the next day (an all-downhill hike that brought us the first overcast of the trip), the two-day drive home, and the next four days at work. Not long after getting home, I Googled Cymbalta side effects, and found insomnia on the list. I threw away the remaining pills, and restful sleep slowly came back, but without normal speech. That Friday, a week after finishing the Olympic hike, I went to Palo Alto, California, for the second opinion from the neurologists at Stanford. The neurologist who initially saw me was a resident or student intern, and after methodically going through the patient history, she left the room to confer with the attending neurologist. That doctor soon returned with the resident, and then a parade of different neurologists and residents came through the room. Tellingly, one neurologist had me take off my shirt to point out twitching muscles on my chest, some of which I didn't even feel. More disturbing was two different neurologists, independently of each other, telling me that they thought I had "a motor neuron disease like ALS." Not MS or Lyme disease or some other neurological disorder that could produce symptoms like mine—and which were at least partly treatable—but ALS, which still has no really effective treatment. Before they could give me a formal diagnosis, however, they needed to run several tests to eliminate the other possibilities. We did one of them the next day, and scheduled the rest, along with a follow-up appointment, for two weeks hence.

Until one has an actual diagnosis (and frequently after), hope that something else is at play springs eternal, the mind grasping at the smallest shreds of evidence supporting an alternate explanation. That was the case for me as I drove to Palo Alto the day before the follow-up visit. I stopped at my optometrist and asked him to look for the telltale signs of MS in my eyes; when he'd finished his exam, he answered with a hesitant maybe. He had not said no, though, so I continued driving, with somewhat improved spirits. Still, I knew that the neurologist appointment was on Friday the thirteenth, which seemed to be another bad omen, even if I didn't suffer from triskaidekaphobia (fear of the number thirteen). Come the day of the appointment, the morning was emotional, as the appointment—and its potential new reality—approached. I knew there were some other tests still to do, but I also knew that the doctor was closing in on a diagnosis no one wants to hear. Lunch with a friend provided a welcome distraction, and then she and I walked to the doctor's office. There, the neurologist spent the first part of the time discussing results from the recent tests, all of which were negative. I knew there were still more tests to do to eliminate other possible causes, so I was expecting the doctor to schedule them (hope springs evermore eternal). Instead, she surprised me with her diagnosis: ALS (although she did schedule the remaining tests, she told me they were unlikely to cause my symptoms). The rest of the appointment was a blur of questions, discussion, and tears—as well as dashed hopes.

I remember that horrible day well, with three things especially standing out. The first was the puzzling expression on the doctor's face once she began telling me her diagnosis. It was probably her attempt to be empathetic, but it looked to me more like the beginning of a smile. I knew it was not malevolent, but it definitely seemed out of place given the awful news she was delivering. The second thing was the lonely trip to a different friend's house for the night. My friend was herself a guest in someone else's home, so I had made plans to stay with a buddy in Oakland that night. With diagnosis in hand, I made one of the loneliest commutes of my life going across the bay. Lastly, when my buddy's wife picked me up from the train station, I couldn't stop crying long enough to tell her why I was upset. I eventually did, and we passed the first evening of my new reality cooking and eating comfort food and generally trying to forget the day's events. It truly was the worst day in my life.

The tears flowed again the next day, when I flew to St. Louis to spend a few days with my folks. Together, we began the process of dealing with the diagnosis. I was single at the time (never married), so they were my closest family, and they reassured me that they would care for me when the time came. We discussed other things that seemed important at the time, but the most important thing we did was just be together, knowing that we would get through it together, as families have always done. Not all families are that way, but I have been blessed with one that is. It would be a source of strength and love that would come to be a salvation for me, as the rest of this book will illustrate. Similarly, the parks would come to play a pivotal role in coming to grips with my new reality, as I sought—and received—solace in them. Life was far from being over for me, just as it's far from being over for the national parks in the face of global warming. The new realities, though, are increasingly removed from the previous, and just like my early symptoms, the first signs of impending changes for the parks were subtle and easily missed. But as with my personal story, it won't be long before the changes become more visible and damaging.

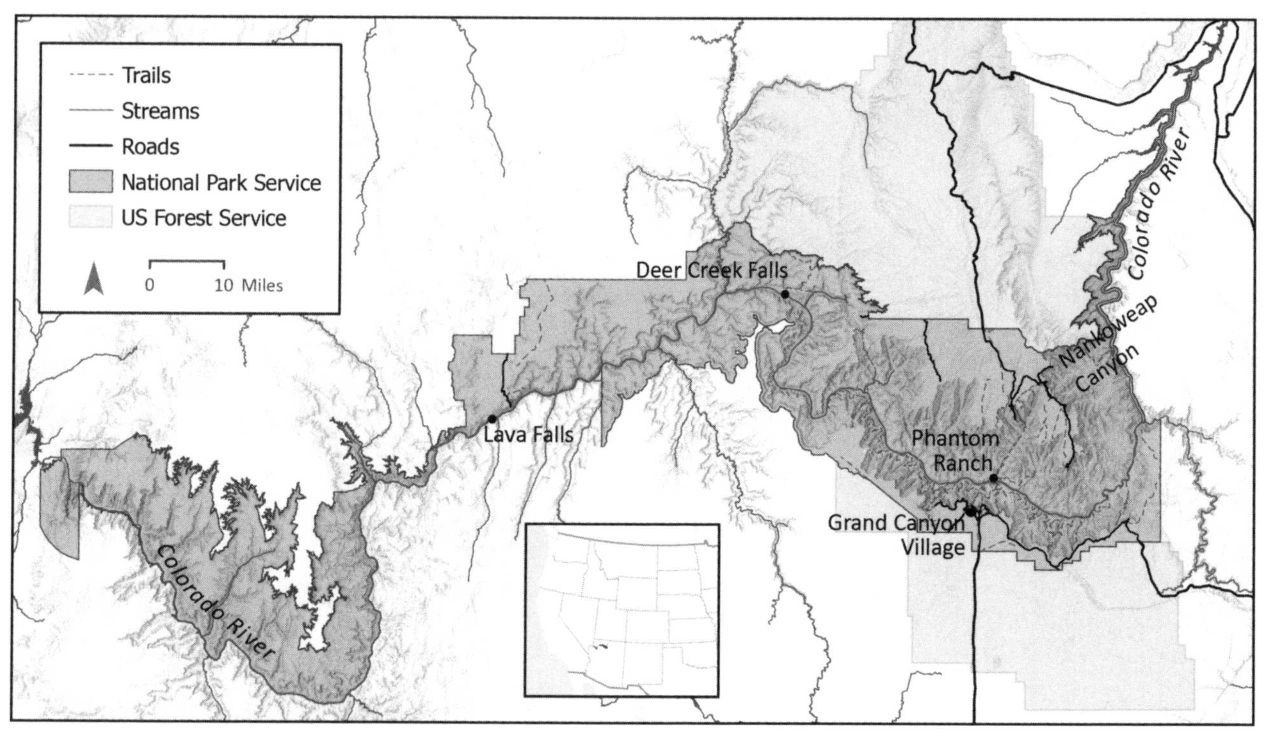

MAP 2 Map of Grand Canyon National Park. *Sources*: NPS, USGS, Natural Earth.

Chapter Two

GRAND CANYON NATIONAL PARK

The Grand Canyon is all about space: the yawning chasm that one looks down upon from above, the heights one looks up at from river level. A mile deep, up to 18 miles wide, and a whopping 277 river miles in length, the canyon overwhelms us with vastness. It is all about time as well—the eons represented in its layers, the time it takes to grasp its age and space. More than a third of Earth's geologic past is present in its layers, while it takes a minimum of a week to float the river miles preserved in the park.[1] Finally, the canyon is all about life too: the six life zones in the park represent everything from Mexican deserts to subalpine forests, and they reflect the wildness that animals impart to the landscape. An extraordinary diversity of flora and fauna call the canyon home, from 1,750 plant species (4 of them endemic) to 373 bird species, 91 mammal species (22 of them bats!), 58 species of amphibians and reptiles, and 5 native fish species. Of these, 20 animals are endemic, including the Grand Canyon rattlesnake. The canyon is truly a world treasure.[2]

Seven months after my diagnosis, I embarked upon a two-week raft trip through the canyon with a commercial rafting company. Two of my brothers joined me for part of the float: my twin, Jim, for the first half, and my younger brother Paul for the second half, switching out at Phantom Ranch. It was to be a convivial odyssey, trading the worries and concerns of our regular lives for incredible sightseeing, exciting rapids, and relaxing camp life. The trip got off to an inauspicious start, however, with a late April snowstorm in Flagstaff, Arizona, where we met up with the group leader the night before we launched. She assured the group that there was no snow where we were going, because it was significantly lower in elevation than the city. As we drove to the put-in at Lees Ferry the next day, we quickly dropped out of the snow and into the desert world that would embrace us for the next two weeks. It was warm and sunny, which boded well for the trip—in everyone's mind, anyway.

Once we arrived at Lees Ferry and met the other guides, we began preparing for the trip. The first day of a river trip is usually filled with lessons from the guides showing their passengers how to set up tents, how to maintain personal hygiene, and how to protect the environment we're passing through (especially regarding human waste). It wasn't long before we climbed into the rafts. That's when the group got to see what my physical limitations were, for I had real trouble getting in. When I sat on the raft's outer tube and tried to swing my legs over and get in the raft, I almost fell into the river, unable to get my center of gravity shifted so I would land in the raft. With Jim's help, I was successful on my third attempt, but Erica, the group leader, and the other guides had seen the unfortunate event.

That evening, after an enjoyable first day, Erica took Jim and me aside to discuss the situation. She had been told that I have ALS, and that my doctor approved my going on the trip, but my physical limitations were now obvious. I had actually met with her supervisor a few months earlier to tell them about my illness and had deliberately asked to shake her boss's hand to demonstrate that my grip remained strong (and, therefore, that I could hold on to the raft in rapids), but my balance had deteriorated since then. She asked if I would be willing to take a swim test without really telling us what that entailed, and I said yes. I dearly wanted to take the trip, so I figured I should express my willingness to do the test, whatever it was. She nodded, said she still had "concerns," and left it at that. She never did have me take the test, but I understood what she was thinking.

The difficulties I was having with balance and walking are a typical part of the disease progression. It's easy to see why, because the two require the coordinated movement of dozens of muscles—many of which were weakening—from the neck to the toes. I had begun falling shortly after the diagnosis, but those early falls were more flukes than a pattern. By the time of the river trip, though, they were happening so regularly that I was using trekking poles as canes whenever I went outside. I was okay inside, where I could generally hold on to or lean against furniture, countertops, and walls. Still, I had taken some serious falls, including one at a friend's house that resulted in my needing eight staples in my scalp to close the wound. The worst fall was just two days before the float trip, when I fell down three steps.

I knew I would play it safe on the trip, using the poles at all times and staying in camp when the other rafters went on hikes. Even though I still fell occasionally, the landings were softer, almost always on soft sand, and I completed the trip without incident. However, that conversation with Erica stayed with me for the first half of the trip. It created an air of uncertainty and made me fear that she would remove me from the trip at Phantom Ranch. Not until the day before we arrived there would we discuss the matter again, with a very different—and much more agreeable—outcome.

From the second day on, a more pleasant routine set in, consisting of coffee and breakfast, breaking down camp, floating, lunch at some beautiful beach, more floating, setting up camp, happy hour, hearty dinner, and mellowing out as the shadows lengthened. I had chosen an outfitter that eschews motors, so we could enjoy the fabulous scenery at the river's pace. Other companies use a motor to travel through the Colorado's many calm stretches, shaving a week from our schedule, but producing a more rushed experience. Our slower pace magnified the relaxing daily routine, allowing us to savor the scene and hear the canyon wren. Not a day passed without hearing the Colorado Plateau's serenade, multiple times. It was a moving performance, the scenes and songs.

Floating down from Lees Ferry, we watched the canyon grow, deeper and deeper. The rock layers that form the canyon gradually revealed themselves as the river cuts down through them. These are the horizontal stripes in the canyon walls one sees so prominently from the canyon rims. Each one represents millions of years of geologic deposition—collectively, 300 million years, ending about 250 million years ago. The Kaibab Limestone, the Toroweap Formation, the Coconino Sandstone. Mile after mile, layer after layer, color after color. First white, then mauve, red, and green, all with an overture of rust, the prevailing color. The harder sandstones and limestones form cliffs soaring practically into infinity, while the softer shales form more gentle slopes and benches (flat areas where the shale has eroded away, leaving the top of the more resistant layer below). The Hermit Shale, the Redwall Limestone, the Muav Limestone. The layers are tall, each one hundreds of feet—collectively, 2,400 to 5,000 feet. Deeper and deeper the river dives, first a few hundred feet, then a few thousand, eventually a vertical mile. The canyon rims are easily visible at first, but they gradually become hidden behind the cliffs at river's edge, eventually to be only rarely seen. The Bright Angel Shale, the Tapeats Sandstone, then, finally, the Vishnu Schist, the basement rock in the canyon. At 1.7 billion years old, it's a third of the earth's age.[3] The hardest of all the layers, the schist resists erosion, forming narrow canyons within the canyon, its walls often rising right out of the river. The schist comes and goes as we float downstream, providing stretches of rockbound intimacy that contrast the more open areas.

The canyon dwarfs the senses, providing no effective scale of comparison. During our trip, we passed the days in wonderment, talking with each other as people do, but never forgetting to fill our minds with the glories around us. Our eyes reached for the sky, following the layers upward. Returning to ground level, our eyes followed the river's ever-changing panorama of cliff, water, sky, and life. Perpetually, the canyon's immensity alternates with intimacy, a riverside spring one moment, a peek of the rim through a break in the Redwall Limestone the next. Vasey's Paradise, a patch of greenery around a spring, gives us a contrast with the aridity surrounding us, then a look down-canyon reveals the rusty ramparts enfolding the seemingly tiny but endless stream. The river's handiwork similarly alternates between grand and life size, scouring out the layers of millennia, but polishing ornate flutes and scallops in the Vishnu Schist, much like the water's movement immortalized in stone. Life exhibits a variety of shapes and sizes, from agaves sending up their twenty-foot stalk of flowers, to prickly pears squatting on the ground, both plants blooming yellow. Bighorn sheep call the canyon home, but so do scorpions; both species command our attention. This canyon truly is grand, on all scales and measures, on all senses, from intimate to immense, impossible to capture in mere words.

The third night, we camped at a place called Nankoweap, across from a gigantic cliff of Redwall Limestone and other layers of stone accompaniment (see plate 10). Nankoweap would come to be one of my favorite camps on the trip, largely because of its setting and the relaxing time we spent there. The Redwall cliff has an equal on the Nankoweap side of the river, and those walls continue parading downstream and out of sight. Because the canyon in that direction is straight, one gets a fine view of the rust colonnades guarding the river to the end of sight and, seemingly, time. My twin, Jim, and Bob Graves, a fellow floater from Michigan who ended up helping me as much as my brothers did, spent the evening at river's edge, watching the setting sun work its way up the cliff face. It was a fine evening made memorable not because we did anything extreme, as is so popular these days, but because we let nature perform while we took in the show. Sure, we were getting our thrills shooting the rapids, but as anyone who has floated the Colorado knows, rapids account for less than 10 percent of the river's mileage in the canyon. The other 90 percent consists of long stretches of calm or gently moving water, and even in the rapids, commercial tours have to play it safe. There was little unnecessary thrill seeking on our trip, which was a good thing. The beauty and tranquility of Nankoweap were ours to embrace.

The next morning, Bob and most of the group went on a hike up to some prehistoric granaries high above the campsite. Ancestral Puebloans built the granaries to store the corn and other crops they raised for later use. We could actually see the granaries from camp, but the hikers were able to get right next to them and look inside. Granaries are simple structures, consisting of rocks mortared into a wall, enclosing an alcove in a cliff face and creating a space of one or two cubic yards. Scattered throughout the Southwest, they were usually built under an overhanging cliff (to keep the food dry) and were sealed (to keep rodents out) until the food was needed. In my travels through the Southwest, I had seen other granaries numerous times and understood the thrill of examining them up close. Sometimes they even have remnant corn cobs—up to seven or eight hundred years old. The arid climate on the Colorado Plateau means that organic materials take a long time to decay (sometimes a really, *really* long time). That people actually lived at Nankoweap made the place all the more meaningful to me, giving it significance beyond its beauty: this was *home* for someone in the past. And why not? The place had year-round water, there was sufficient flat ground for farming (Nankoweap has more of that precious commodity than most of our campsites, at least for one family group), and there was abundant security.

Like many places in the Southwest, though, Nankoweap was abandoned hundreds of years ago. All throughout the Colorado Plateau, Ancestral Puebloans left their homes and moved south, some resettling along the Rio Grande in today's New Mexico, and at other places nearby. The reasons for their departure remain a source of significant debate, but one clue comes from the very structures they left behind. In larger structures where they needed to use logs for beams and other supports, researchers can remove a core sample of wood (much like an ice core) to study the growth rings. As we learned in childhood, trees in temperate climates put on growth rings every year, and the width of those rings generally corresponds to the relative abundance of soil moisture fed by rain in that growing season. By determining when the tree died and counting back along the growth rings, the scientist can date them. Next, the researcher can calibrate the ring widths to specific amounts of precipitation by comparing them to the known precipitation of recent years (as recorded by nearby weather stations, which is the "instrument" record). Finally, the researcher can go further back in time by cross-dating the overlapping ring widths of a dated tree with those of older trees from the same area. In this way, scientists can go back thousands of years where trees are available—like the arid, but peopled Colorado Plateau—to get a good idea of the changes in the moisture regime there over time. Also, by matching a core sample taken from a structural support to nearby dated sequences, we can tell when the tree died and, therefore, the approximate year of the structure's construction.[4]

In mountainous areas, tree rings also hold clues to the warmth of past climates. In the interior American West, mountainsides typically have two treelines, a lower one below which the climate is too dry for trees, and an upper one above which conditions are too cold for them. By gathering core samples from both treelines and then cross-dating them to each other and calibrating them to the local instrument record, scientists can tell how warm (from the upper treeline) and wet (lower treeline) the climate was in years past, with a surprising degree of accuracy. Researchers have done numerous studies like this on the Colorado Plateau, including at least one that took core samples from both sides of the Grand Canyon. Collectively, the tree rings from the Colorado Plateau paint a picture of a region that is relatively warm and dry, but is quite variable year to year, decade to decade, and century to century. Moisture can double or halve one year to the next, and the average annual temperature can vary yearly by as much as 5 degrees (which is a significant amount). The place can see its moisture regime flip in one year to a drier one that lasts decades. Most importantly for this discussion, though, is that the Southwest is perpetually on the edge of drought—long ones, short ones, mild ones, and severe ones. Most likely, it was one such drought that sent the Nankoweap granary builders packing, the site's limitations eventually overcoming its advantages. Throughout the Southwest, people have abandoned their prehistoric settlements—some quite large, like those at Mesa Verde and Chaco Canyon—when drought struck. The region's dry climate could only support so many people in good years, and drought forced a reckoning of some kind upon its residents.[5]

That's not all the tree rings tell us. There is one time period in the Grand Canyon study that stands out for its excessively hot climate, the warmest in its 1,425-year period of record: the time from 1950 to 2010.[6] The heat has magnified since then, causing a 19 percent decline in the Colorado River's volume. While previous droughts of similar intensity were caused by reduced precipitation, higher-than-normal temperatures—and the added evaporation that comes with them—are the culprits here. Around a third of this loss in river flow (one-sixth to one-half, with one-third being the average) is directly attributable to anthropogenic warming, a conclusion reached by two different groups of researchers analyzing the issue.[7] Doing the math, 7 percent of the Colorado's annual flow is gone, evaporated into the hot desert air by the heat we have caused. If that doesn't seem like much, it's equivalent to a fourth of the water that California draws from the river, or 38 percent of Arizona's withdrawal. Adding the Californians and Arizonans whose drinking water comes from the Colorado River to the people of Las Vegas, Nevada, who also subsist almost entirely on its water, we find that more than thirty million people depend on this desert river. Combine this with the irrigation water supplied by the Colorado to the farm fields of California's Imperial Valley, which grows most of the nation's winter produce, and it's clear that a seemingly small reduction in this river's flow can have far-reaching consequences.

If we continue business as usual, flow reductions will only get worse, with the projected losses growing to 20 percent by the middle of the century, and 35 percent by century's end; losses could even be as bad as 30 percent by midcentury and 55 percent by the end.[8] Moreover, if precipitation drops, as predicted in some climate models, there will be even less runoff, so little that another group of scientists project a 15 percent reduction in flow for every 5 percent reduction in rain- and snowfall (and, it should be noted that although other climate models project modest increases in rainfall, that has yet to occur).[9] Such steep reductions prompt an important question: If people with such modest demands on their environment as the Nankoweap granary builders were chased away by a changing climate, how will we fare with our much larger population and associated environmental demands if these projections become reality? We may have more options at our disposal to cope with climate change (like air conditioning and tapping groundwater), but as we'll see at the end of the river trip, Lake Mead, the country's largest reservoir, is less than half full, and Lake Powell is no better.[10] In other words, there is only so much water out there, and it's declining in abundance. We would do well to contemplate the lessons that Nankoweap and the many other southwestern parks preserving Ancestral Puebloan ruins provide about living within nature's limits. No society is immune from the constraints of nature, especially in an already limited environment. Our own day of reckoning may soon arrive, driven largely by our own actions.

Returning to the physical affliction that was so obvious on the first day of the trip, let's pause for a brief discussion of ALS's history and prevalence. It was first described by Scottish physician Charles Bell in 1824, and fifty years later, French neurologist Jean-Martin Charcot linked the symptoms to the underlying neuronal degradation. He also gave the disease its contemporary name: amyotrophic, which means "no muscle nourishment," referring to the loss of neuronal signals to the muscles; lateral, which refers to the areas in the spinal cord where the affected motor neurons are located; and sclerosis, or the death of the motor neurons. While the disease is the third-most-common neurodegenerative disease (after Alzheimer's and Parkinson's diseases), it's a relatively uncommon affliction. Worldwide, about 4.5 people per 100,000 have the disease, with 1.9 per 100,000 diagnosed with it yearly. Diagnosis consists of eliminating all diseases that have similar symptoms, there being no positive diagnostic test. While two genes have been shown to cause ALS, only about 5–8 percent of individuals with it have one of them. Researchers believe there may be other genes at play, potentially in all disease sufferers, but more research is needed to confirm that hypothesis. With no effective treatment to slow the disease progression, physicians instead treat the symptoms, prescribing muscle relaxants for tight muscles, for example. From diagnosis, 80 percent succumb to ALS within five years, while 10 percent live longer than ten years—but the disease is 100 percent fatal, its progression unstoppable, the affliction all consuming.[11]

Not so long before this trip, I would have happily joined the group on their hike to the granaries. I was an avid backcountry enthusiast, regularly taking to the trails and wilderness waterways to explore whatever national park I was working in, along with many nearby parks, wilderness areas, and national forests. It was an unusual year that I did not put five hundred miles on my hiking boots, with several dozen more miles on my cross-country skis and in my canoe. In the months leading up to this trip, though, I saw my ability to walk and ski gradually and inexorably diminish. Skiing was the first to go, as I found out two months before this trip. I had not been on my skis yet that winter, and was already beginning to have trouble walking. Visiting Yellowstone friends that February, I went out on an easy cross-country skiing trail, and did okay on the gently uphill route away from the trailhead. Turning around for the ski back, though, I quickly began to have trouble with the simplest downhill stretches, repeatedly falling into the soft snow. After a half dozen attempts, I gave up, took my skis off, and walked the rest of the way back, carrying my skis. Sliding on skis demands that one's muscular coordination be close to perfect, a level of muscular finesse that I clearly no longer had. So, skiing became the first activity I enjoyed to be taken from me by ALS. There would be many, many more.

Indeed, my single most cherished pastime, hiking, would be next. Walking doesn't demand the same level of fitness as skiing, so I was able to hike for a few more months. Still, the trend in my hiking mileage and level of rigor of the hikes I took was hard to ignore: shorter and shorter hikes, easier and easier trails. One that stands out in hindsight is a March hike to a place called Hite Cove outside of Yosemite, a hike I did every spring to admire the profusion of California poppies and other wildflowers. The four-mile hike to the cove crosses some steep slopes and scrambles over one or two rock outcrops. Previously, I had bounded right through those tricky areas, but not that time. Even though I was using trekking poles, I still was so unsure of my footing that I got down on my hands and knees and crept through the rocky zones. I suppose I should not have been surprised at my struggle that day, for two weeks prior I had taken a similar hike and found myself to be much slower than my two hiking partners, especially on the downhill stretches. But the disease progression was still new to me (notwithstanding the skiing incident), and it hadn't deprived me of something as critical and basic as walking. Furthermore, it's human nature to hope that you'll be able to hold on to favorite pastimes in the face of life-altering changes. For these reasons, the difficulties I experienced hiking the Hite Cove Trail that day left me incredulous. Was the disease really moving that fast? Would I have to give up hiking, and the freedom it gave me to explore beautiful places, so soon? Was I really moving that quickly toward disability?'

At Nankoweap, it was time to move on down the river once the hikers returned. After eight more miles of colossal beauty, we arrived at the confluence of the Little Colorado with the larger river. When it's not muddy with thunderstorm runoff, the tributary is an otherworldly turquoise blue, the result of flowing through limestone upstream. It's so brilliant that it almost glows, much like a snowy landscape on a clear sky dawn. Instead of snow, though, white sand accentuates the brightness. Contrasting the brilliance on this day was the cloudless desert sky, as deeply blue as the Little Colorado was light. Bridging the blues were the canyon walls, three thousand feet of red and olive and taupe. Adding hints of green in the foreground were scattered patches of tall grass, reveling in the rare moisturized soil. It was also a landscape of angles and texture, from the soft, almost creamy streambed of imagination to the soaring cliffs that are at once vertical and horizontal—one direction of cliff, the other direction of layers. It is yet another remarkable landscape, one of many in a place impossible to describe in words.

We spent almost an hour there, admiring the color—and wondering why anyone would want to sully that sacred place with a gondola tramway shuttling up to ten thousand people a day down from the nearby canyon rim. A Phoenix company called Confluence Partners LLC wanted to build the "Grand Canyon Escalade" on 420 acres of Navajo Nation land on the canyon's East Rim. The project would have included an IMAX theater, shops, hotels, and a 1.5-mile tramway that would have whisked people to the confluence in a mere ten minutes. There, the tram riders would have found two more restaurants, a gift shop, a sculpture garden, an amphitheater, and a covered walkway leading to an overlook with signs describing what had been sacred, all at the edge of the two rivers. Promoters extolled the benefits of the tramway, saying it would enable millions of people (most of whom would not otherwise be able to get there) to experience the inner canyon and the Little Colorado, and promising the tribe its share of the millions of dollars spent there and on the rim. Promoters also pointed to the Skywalk on the canyon's South Rim, the horseshoe-shaped walkway suspended over the canyon abyss, as an example of a supposedly benign development that is bringing the Hualapai Nation its own millions. Critics responded that it would violate a place that is sacred to the tribe, and that it would be an unconscionable intrusion of development and commercialism into the canyon. Lost in the rhetoric was the fact that any able-bodied person—and a fair number of partly disabled people such as myself—already have access to the inner canyon on rafts and dories, without the unconscionable intrusion of greed in the natural temple. Truly, it's hard to imagine a more naked attempt to make a buck at the expense of the public and nature, masquerading behind booster rhetoric. Perhaps this is why Confluence Partners was a limited liability corporation: they didn't want to be credited with such a god-awful idea (but they would've been laughing about it all the way to the bank). Thankfully, the Navajo Nation Council voted 16–2 against the project in late 2017, killing it, at least for now (bad ideas have a way of resurfacing, especially when there is money involved).[12]

If bad ideas don't reappear in the same form, they can be reincarnated as a different money-making travesty, as this very stretch of river bears witness. It was not a reincarnation of the Grand Canyon Escalade (give it more time), but rather an earlier proposal that, proverbially, came back in the form of the tramway. In the early 1960s, the Bureau of Reclamation (USBR) proposed putting two dams in the canyon, one just twenty-two miles upstream from the confluence with the Little Colorado, and the other far downstream, just before our boat takeout at Lake Mead. The upper one, known as the Marble Canyon Dam, would have been 310 feet tall, and would have flooded Vasey's Paradise and the Redwall Cavern, a gigantic overhang we'd seen. The lower one, known as the Bridge Canyon Dam, would have been a giant, standing 740 feet tall and creating a reservoir ninety miles long, backing forty miles into Grand Canyon National Monument/Park (the monument, now part of the national park, extends downstream from the park). It would have inundated Havasu Creek (another turquoise creek with several tall waterfalls upstream) and Lava Falls (perhaps the most dangerous rapid on the Colorado). The spoils from exploratory drill holes into the canyon walls at both damsites can still be seen. At one point, the proposal for Marble Canyon included a thirty-eight-mile, thirty-six-foot-diameter pipeline that would have taken 90 percent of the river's flow from Marble Canyon to a power-generation site west of the park. The power plant would have produced almost twice as much electricity as Hoover Dam downstream, thanks to a 1,300-foot hydraulic head. The two dams would have only been "cash register" structures, generating electricity but providing no flood control or irrigation water. Instead, the funds from the sale of their power would have been used to finance other reclamation and pipeline projects elsewhere in Arizona.[13]

By the late 1960s, public outcry on the dam proposals was reaching a fever pitch, with discussion focusing on geologic problems with the Marble Canyon Dam site and impacts on the park with the Bridge Canyon Dam. The political climate was shifting too, with the Central Arizona Project (CAP), which would deliver Colorado River water (diverted downstream of the canyon) to Phoenix and Tucson, under consideration at the same time. Increasingly, it appeared that public opposition, stoked by the conservation community, would derail both the dams and the CAP, so boosters of the projects were considering dropping the more contested dams. A decade earlier, the Sierra Club, led by David Brower, had been a major player in a similar controversy, one that would come to have a bearing on this one. Brower had been one of the architects of the compromise to keep the Echo Park Dam from being built in Dinosaur National Monument in Utah and Colorado. The deal was that, if conservationists would not oppose a dam in Glen Canyon (on the Colorado River just upstream from the Grand Canyon), the USBR would withdraw the Echo Park Dam. Both sides kept their end of the bargain, but before the dam was complete, Brower and some of his associates floated through Glen Canyon. They found it to be a place of exquisite beauty accessible to almost anyone, for there were no rapids there. They realized they'd made a mistake in agreeing to the compromise, so when the USBR made public its plans to put two dams in the Grand Canyon, Brower sprang into action. Galvanizing his organization's members and networking with other conservation groups, he took on the USBR. One of his tactics was to take out full-page ads in the *New York Times* and *Washington Post*, hoping to sway public opinion against the dams. The most provoking of these parroted the USBR's claims that the Bridge Canyon Dam would enable more people to experience the inner canyon by providing motorboat opportunities on the reservoir. In bold, all-caps lettering, the ad shouted, "SHOULD WE ALSO FLOOD THE SISTINE CHAPEL SO TOURISTS CAN GET NEARER THE CEILING?" The ad went on to urge readers to write Secretary of the Interior Stewart Udall in opposition to the dams. Hundreds of people responded, and with many other conservation organizations, newspapers, magazines, and politicians allied against the dams, momentum was turning in favor of the conservationists.[14]

In 1968, the twin dam proposals died as a combined result of the opposition, shifting political climate, and technical issues with the Marble Canyon Dam site. In 1969, President Lyndon Johnson established Marble Canyon National Monument, which was incorporated into Grand Canyon National Park six years later, along with Grand Canyon National Monument. With both reservoir sites included in the expanded national park, the fight over reclamation in the canyon was over—but not the ongoing threats to the canyon's integrity, as the tramway illustrates.[15] Someday, someone else will come up with another grand scheme to denigrate the canyon to make money. I hope we'll have the sense to do as President Theodore Roosevelt urged us a century ago, when he visited the canyon: "In the Grand Canyon, Arizona has a natural wonder which is in kind absolutely unparalleled throughout the rest of the world. I want to ask you to keep this great wonder of nature as it now is. I hope you will not have a building of any kind, not a summer cottage, a hotel or anything else, to mar the wonderful grandeur, the sublimity, the great loneliness and beauty of the canyon. Leave it as it is. You cannot improve on it. The ages have been at work on it, and man can only mar it."[16]

Moving on from the Little Colorado, another twenty-six miles of floating brought us to Phantom Ranch, a small, historic collection of rustic cabins and a canteen with free iced lemonade for thirsty hikers and boaters. The ranch is on several hiking trails that go to the canyon rims, so it's the changeover point for river trips. My twin, Jim, traded places with my younger brother Paul, who would be with me the rest of the trip. I walked with Jim to the canteen (it was a well-maintained, flat trail, so I had no problem with it), and said goodbye to him. I broke down and sobbed in public there, it being much more difficult to part ways with any loved one now. An odd symptom of ALS, known as pseudobulbar affect, makes tears flow more readily than before for me (the affect causes other sufferers to laugh more readily than cry). Fraternal twins like us don't always develop the innate, tight bonds that identical twins do, but since fraternal twins are still constant companions for their first eighteen years, they often develop similar linkages. Jim and I certainly have them, so he helped me compose myself, and we said our final goodbyes. It would only be four months till we'd be together again.

Climbing back in the rafts with Paul (who was as gracious as Jim), we introduced him to the river with a series of the river's biggest rapids: Horn Creek, Granite, Hermit, and Crystal, all of them ranking at least a seven on the canyon's ten-point scale (ten being the most difficult). We also introduced him to the community we had become. Perhaps the best illustration of this came a day or two before we arrived at Phantom Ranch, when I asked Erica if she was going to make me leave the trip at the ranch. She hadn't asked me to take the swim test, but also hadn't given me any other indication of her thinking. In answer to my question, she said something like, "Are you kidding? I'd have a mutiny on my hands if I did. Everyone wants to see you finish the trip." I was pleased but surprised; I suppose they could all see the effort it took me to go through the day, as well as the enthusiasm I had for the beauty surrounding us. They had become accustomed to letting my brother go first through the dinner line to prepare my plate, understanding that my weakening tongue and throat muscles made eating a slow affair. I would begin eating before any of them, but I would usually be the last one done. In the more dangerous rapids, Paul and Bob would move in close to me, one on each side, to help keep me in the raft should we hit a wave wrong. Not only did I stay in the rafts for the duration, but no one else took an unexpected swim. Quite unintentionally, I had become the glue holding the group together, as one of the passengers put it.

The last of the wild rapids in this stretch, Crystal, was greatly enhanced by a huge storm in December of 1966. Over three days, fourteen inches of rain fell on the North Rim (just three inches short of its annual average), with two inches at Phantom Ranch (a fourth of its yearly total). Debris flows occurred throughout the canyon, including in the Crystal Creek drainage, where the creek was the size of a small river. The Crystal debris flow dumped so much sediment, rocks, and boulders that the Colorado was pushed against the far side, constricted to just 20 percent of its former channel. The material also acted as a dam, backing the river up and heightening its drop at Crystal Creek. Subsequent floods have carried some of the material away, but Crystal remains a rapid that demands boater vigilance. Today, the river is split into two channels, with a gravel bar between them; should a hapless boater get stranded on the island, the only rescue available is by helicopter, the river dynamics preventing safe retrieval by boat. The rapid remains a vivid illustration of nature's power—one of hundreds of such illustrations in that grand place.[17]

The storm that created the rapid was probably caused by an atmospheric river, a plume of moisture originating far out over the Pacific Ocean, often near Hawaii. Such rivers are a wintertime phenomenon, but they usually deliver rain instead of snow, given their subtropical origin. They are examples of extreme weather events, paralleled in summertime by strong thunderstorms. Summer thunderstorms may already be increasing in strength and frequency due to global warming and are expected to increase further as warming intensifies.[18] Grand Canyon has already warmed by as much as 4 degrees, and the decade 2001–2010 was the warmest in the 110-year instrumental record (as noted in the introduction).[19] Recent temperatures there have been so warm that they are at the extreme warm end of the park's historic weather, hotter than 99 percent of the various measures of warmth (i.e., annual average temperature, minimum temperature of the coldest month, average summer temperature). Grand Canyon is not alone in its warming, for almost all other parks in the desert Southwest (and *all* of them in Arizona and New Mexico) were similarly hot.[20] As a result, spring is arriving earlier, and the growing season in the Southwest has lengthened by about two weeks since 1950.[21]

Some plants and animals may already be responding to the warming climate. Elegant trogons, birds previously known to occur only in the borderlands of southeastern Arizona, were observed nesting at both units of Montezuma Castle National Monument in 2008 and 2009 (the monument is about halfway between Arizona's southern border and Grand Canyon). Warmer winter and spring weather might be luring the colorful bird north to places that were previously too cold.[22] Forests on the rims are likely responding to the heat in less salubrious ways, by dying from bark beetles tunneling into the cambium (growing tissue) of stressed trees and girdling them, or in wildfires. About 18 percent of forests in the Southwest perished between 1998 and 2008 to one or both of these causes. Also, the number of wildfires over a thousand acres in size in the high-elevation forests of Arizona and New Mexico has increased significantly since 1984. This includes the forests on the North Rim, which is about a thousand feet higher than the South Rim; the additional elevation brings more precipitation, especially snow, which in turn supports a mixed conifer forest that includes spruce and fir, trees more typical of the Canadian Rockies. Other factors are likely contributing to the increased fire activity, such as fire suppression for several decades before the 1960s. Today, the NPS allows some fires to burn when conditions will keep the fire from getting out of control. This replicates fire's role in the forest ecosystem, burning the needles and branches that accumulate on the forest floor, and thinning the seedling trees that have sprouted since the last fire. However, prior to the 1960s, park managers (and USFS managers throughout the Southwest) suppressed all fires, leading to a build-up of fire fuels and thickets of saplings. The net effect of this suppression is that forests in the Southwest are, in general, more flammable than they were a hundred years ago. Even accounting for the effects of suppression and other complicating factors, though, the chances of the large fire increase occurring in the absence of human-caused climate alterations is just 2 percent.[23]

Trees are dying in the region's mid-elevation forests as well, with the case of the piñon pine (*Pinus edulis*) being particularly revealing because its decline points directly to climate change. This tree is a signature component of mid-elevation forests in the Southwest, usually found with juniper trees, and the nuts it produces are edible, nutritious, and sought after by humans and wild animals alike.[24] The pines depend more on winter snowfall than on summer thunderstorms, the two main sources of precipitation on most of the Colorado Plateau. Winter storms, whose snowfall gradually melts and soaks into the soil, primarily benefit woody plants with deep taproots like the pine, while summer thunderstorms primarily benefit annuals and other plants with branching root systems that can sop up rain quickly. With winters shortening and summers getting hotter, the pines are likely experiencing more drought stress than they can tolerate. Even though there are no discernible trends in the amount of precipitation across the Southwest, the trees are dying, even in wetter, higher-elevation sites like Mesa Verde National Park and Bandelier National Monument, both of which saw most of their piñon-juniper forests consumed by recent wildfires. In Mesa Verde, a total of six large fires burned in recent years; the acreage burned has no precedent in at least the last three centuries. Mortality in some sites is near 100 percent, and some 7,500 square miles are affected.[25] Not all populations of the trees are dying, however; piñon pines on the Grand Canyon's South Rim appear to be holding their own, likely because the South Rim receives the most precipitation of any piñon pine woodlands. That may be enough to enable the trees to survive the increased evaporation of our hotter, present-day climate, but it's not clear how much more warming they will be able to withstand.[26] Indeed, there is evidence indicating that piñon cone production has fallen 40 percent across the region since the 1970s, with late summer temperatures being the cause of the decline. The declines correlated with temperature increase: the more that late summer temperatures have increased, the greater the drop in cone production. As it happens, the higher elevations at which the trees grow have seen the highest temperature increases—as with the South Rim.[27] With higher temperatures hampering seedling establishment in other ways too, the outlook for the pine is not rosy.[28]

Indeed, if current trends continue, the piñon pines on the South Rim likely will go the way of their relatives, the precipitation too little for the enhanced evaporation the hotter climate will bring. Furthermore, the piñon won't be the only tree finding the evolving climate too hot and/or dry by the end of the century. By then, temperatures will be 4 to 9 degrees warmer compared to the time period 1960–1979, heat waves will be more common, and the number of days over 95 degrees will more than double, to forty or more each year. Winters will continue to shorten (to thirty days less by midcentury), and the late winter storm track will shift northward, with reductions in snowfall keeping pace. Droughts will become more severe, frequent, and long lasting—some even exceeding fifty-year-long megadroughts in the tree-ring record. Some of the droughts may interact with the periodic climate phenomenon La Niña, which brings dry and warm winters to the Southwest, to produce exceptionally dry winters. Thunderstorms will become stronger, so extreme downpours will come to account for more and more of summertime precipitation, producing more flooding. Winter precipitation will also become more variable, bouncing between extremely wet and extremely dry winters (such as the winter of 2004–2005, the wettest on the Colorado Plateau, and the next winter, the driest). In short, the already arid climate of the Southwest will become much drier, to the extent that one group of researchers call it an "unprecedented fundamental climate shift . . . that falls far outside the contemporary experience of natural and human systems in Western North America."[29]

The consequences for the plants and animals of the region will be severe. Piñon pines don't produce large numbers of cones until they are at least seventy-five years old, and their seeds are only viable for a year or two, factors that limit their ability to move as climate change moves their preferred habitat upward and northward.[30] As the piñon pines succumb, so will three bird species that ornithologists consider "obligates," birds whose specialized needs can only be met by a specific environment. The pinyon jay, gray vireo, and juniper titmouse are all considered piñon-juniper forest obligates, and modeling for their global-warming-influenced future indicates that the park's continued habitat suitability for them is expected to worsen.[31] With snowfall decreasing or even disappearing altogether (as happened in parts of the Colorado Plateau in the winter of 2017–2018), most tree species in the Southwest will find their current range unsuitable for growth. Gambel oak and Douglas fir will be the relative winners, though the two species will still need to move to areas of newly suitable habitat. At the highest elevations (such as the North Rim), the forests reminiscent of Canada will be heated out of existence, replaced with ponderosa pine and Gambel oak.[32] Plant communities typically found together today may dissolve as individual species move in response to the changing conditions. For example, sagebrush may lose its association with rabbitbrush (another common resident of the sagebrush steppe), occurring instead with ocotillo (a Sonoran Desert plant). By century's end, the projected climate on about 55 percent of the western landscape will be incompatible with the vegetation there now, so there will be considerable upheaval occurring in the West.[33]

For these reasons and more, some scientists think of the Southwest as a climate change hot spot, a region that's already on the edge, with native plants and animals already being stressed by a climate that is increasingly too hot and/or dry.[34] It won't take much to push them beyond the brink of their tolerance, as piñon pine may already be demonstrating. It's sad to think of the changes in store for the region, whether they are dead piñon pines, killed by warming winters that no longer deliver the snow they need to survive, or the loss of the unique birds that depend on them. True, other species may take the place of some of the departed, but it still seems like the native communities of plants and animals shouldn't be disrupted, or if disruption is inevitable, that we keep it to a minimum. Indeed, if Theodore Roosevelt were alive today, I suspect he would exhort us to be as vigilant about preventing the worst of climate change as we should be in keeping claptrap development out of the canyon. Moreover, I suspect he would say that we not only owe that to future generations, but also to the strength of our nation, for no country can persist without its inspiring landscapes.

On the river, we floated through the lower canyon, a reach that most park visitors don't see, even from the rim. Those days seem timeless in my memory, a never-ending journey through the most stupendous desert scenery imaginable. Above us soared walls of red and ocher, burnt orange and cinnabar. Above the colonnades we glimpsed esplanades, and above them, higher colonnades. Drifting down from the heavens was the canyon wren's lilt, the lyrics of time as it were, with the rock layers representing millions of years. Twice more, the canyon walls closed in, cloistered by the Vishnu Schist and other metamorphic basement rocks. Scenic intimacy substituted for expansiveness, giving us up-close pictures of schist scallops and swirls, flutes and cups, the rock too hard for the river to push too far aside. Where the walls open up, bighorn sheep were occasionally present, browsing on riverside vegetation or looking at us from a rock pinnacle. Usually in groups of a handful, the sheep seem a fitting resident of the canyon, their elastic hooves giving sure-footedness in that vertical landscape. Well adapted to their environment, desert bighorns can go for days without water and include cacti in their diet, using their horns and hooves to remove the spines. Other riverside attractions included Havasu Creek, its waters as turquoise as the Little Colorado's; Deer Creek Falls, seemingly emerging from solid rock and creating a lovely plunge pool; and a wall of rock art, left by earlier residents of the canyon (see plate 11). The days now seem like a dreamscape, a world of serenity and beauty.

They were days of fun too. By this time in the trip, we had all gotten to know each other fairly well, so bantering between the guides and clients was common. One day, for example, I turned around to watch the receding scenery for a while, which meant I was facing our raft guide, a woman named Heather. Struck by the imposing view, I immediately exclaimed, "Wow!" Without missing a beat, she said, "Thanks, I get that a lot." On another occasion, I laughed so long and hard that I lost part of my breakfast. The evenings were as relaxing as the days were fun, with the group gathering for hors d'oeuvres and drinks before a hearty meal. The various companies that provide river trips through the canyon pride themselves on their backcountry cooking, so we enjoyed lasagna, pork roasts, salmon steaks, and other delicious dinners. (By freezing the food and cooler to a temperature well below 0 degrees and shielding it from the desert heat with the camp mattresses, outfitters can sufficiently cool fresh food and meat for two weeks.) The group continued to help me with daily tasks; one woman cleaned my glasses every evening while another washed and medicated my feet, which had developed a rash after being sunburned. Paul and Bob continued being extraordinarily helpful, setting up my tent and connecting two camp mattresses together so I could turn over more easily at night. This community of wilderness travelers certainly made the trip more enjoyable for me, and for all.

The common denominator in that set of tasks was my weakening fingers. Early in the disease progression, I had had a discomforting glimpse of my future on what would become my last backpacking trip, a three-day hike in the High Sierra just outside of Yosemite with my buddy Jim Roche. It was late in September (just two weeks after my diagnosis), and our first campsite was above nine thousand feet, and cold. Hoping to get my tent set up before I donned warmer clothing, I found my fingers getting colder and colder, and as a result, less and less useful. By the time I finished, I was fumbling as much as I was making progress, and getting into extra clothing was no better. Jim finished with his tent long before me; by the time I finished, he had dinner ready. I ate it hastily and then climbed into my tent, wondering if my fingers would ever recover. As the night passed, they warmed up and their strength—such as it was—gradually returned, but the experience certainly was a peek into my future, and not a happy one.

Over the winter, I found out just how much we take healthy fingers for granted. Mundane tasks of every kind gradually became increasingly difficult. Buttoning shirts was especially hard, even with a tool that helped pull the little things through their tiny holes. One morning I lost my patience with the task and ripped the partially buttoned shirt off, sending some of the buttons flying. The action gave me a primal sense of satisfaction but left me no closer to being dressed. Writing checks to pay bills was no easier; my checkbook became a register of my deteriorating handwriting as I paid the monthly bills. Cooking grew increasingly simplified as I abandoned vegetables and dishes that required extensive preparation. Also, throughout the winter I found myself going through the grieving process, jumping from denial to bargaining to depression and back again. My friend Jodi Bailey, who lived downstairs from me in a one-bedroom apartment, regularly offered her assistance, such as opening my mail. More importantly, she came upstairs whenever she heard me sobbing, offering her shoulder and an embrace. Friends in need, like her, are friends of the highest caliber, indeed.

Yet the winter was notable for more positive things too. My twin, Jim, flew out for a long weekend in Sequoia National Park, where we had lunch sitting on the same fallen sequoia in Log Meadow I had lunched on when I was a seasonal ranger there twenty-three years earlier (giant sequoias are rot resistant, which, when combined with their enormous girth, means they can take centuries to decompose). That was always a special spot for me, a place where I could feel nature's power, an attribute that time had not dampened. Throughout the winter, friends regularly accompanied me on hikes closer to home in Yosemite, whether climbing Liberty Cap (a granite dome near Yosemite Valley) or the Cold Canyon switchbacks (west of the valley), or snowshoeing to the Muir Grove of giant sequoias, back in Sequoia National Park. The mileage of the hikes progressively declined, forcing me to more sedentary pursuits, one of which was on my own deck. I enjoyed two cups of coffee out there on Easter morning, and found deep, rich silence much of the time, broken only by an occasional car on the country road below the house. It was as moving an experience as any church service, at least for me. Nature, especially in the national parks, was proving to be as great a solace as anything I did.

As we floated into infinity on the river, we gradually left the desert of the Colorado Plateau behind and entered a long finger of the Mojave Desert that follows the river up from Nevada. Centered in southeastern California, southern Nevada, and northwestern Arizona, the Mojave is well represented in our national park system, with Joshua Tree and Death Valley National Parks and Mojave National Preserve also preserving portions of this desert. For us, gone were the agaves that end their lives with a twenty-foot-tall stalk of yellow flowers, replaced by an equally tall plant of interest, the Joshua tree. The Mojave's signature plant, the Joshua tree is a member of the yucca family, looking much like a collection of typical yuccas, each at the end of its own branch, all united by a single, gangly trunk. Though I don't remember seeing any of the distinctive trees at river level (they're found a little higher than the river), there were creosote bushes, named for the way the plant's waxy leaves smell of railroad ties, especially after a rain.[35] Waxy, tough leaves are just some of the adaptations to aridity that Mojave plants possess; others are small leaves, hairy leaves, and closing the leaf pores by day and opening them by night, when humidity rises. Such stomata, as these openings known, let carbon dioxide in and oxygen out to the photosynthetic cells; separating the air exchange from the actual photosynthesis is known as Crassulacean acid metabolism. And, as I was lucky to see driving through a winter storm over California's Walker Pass in 1989, plants in the Mojave also need to be able to withstand snow and frost—seeing Joshua trees with snow on them was a unique experience. Scientists believe that such cold snaps, longer than twenty-four hours, are what prevents Sonoran Desert plants from moving northward and upward from their home in southern Arizona.[36] The Mojave is indeed a four-season place of wonder and beauty, one that we were fortunate to witness on our trip.

The Mojave Desert plant adaptations are being put to the test by a double whammy of climate challenges: rising temperatures and falling precipitation. Since 1895, Mojave temperatures have risen by about 3 or 4 degrees, with weeks with an average temperature above 81 degrees (heat waves) increasing from zero to six. In the same time period, the most precious commodity in most deserts—water, or precipitation—has dropped by almost 40 percent.[37] By 2003, the end of one of the droughts that are collectively reducing Mojave precipitation, researchers noticed widespread mortality of woody plants there, though both Joshua trees and creosote bushes appeared to hold their own.[38] Creosote bushes, however, responded to the drought by shedding entire branches, a last-ditch effort by any plant to reduce its water demand. Perennial grasses didn't do so well, disappearing entirely from some locations. Declines in winter precipitation drove these changes, the gentle rains of winter (and occasional snows) failing to soak in and replenish the water that deep-rooted, perennial desert plants draw upon to sustain themselves through the long, hot summer. One group of researchers documenting these changes wrote that the precipitation variability was enough to change the Mojave Desert ecosystem, which is probably not too exaggerated, given the role of moisture as the most influential limiting factor in the Mojave.[39]

Actually, that ecosystem appears to be not just changing, but even possibly collapsing, as evidenced by declining bird populations in the Mojave. In 1908, University of California professor Joseph Grinnell began inventorying sites throughout the state for their bird populations (and, as we'll see in the Yosemite chapter, small mammals). These inventories, which his colleagues continued through the Mojave Desert through 1968, have become invaluable in assessing the impacts of climate change. Repeating Grinnell's transects (inventory lines) today and comparing them to his initial inventory, one group of researchers found a stunning 43 percent decline in the number of bird species at those sites. Only the common raven increased in distribution, and declining precipitation was the cause of all the declines.[40] Birds occupying the warmest and driest sites were more likely to disappear, largely due to their need to be near water to get wet and cool down through evaporative cooling.[41] Smaller birds, which have higher respiration rates than larger birds, were more vulnerable than their larger cousins.[42] Whole food chains were affected, as demonstrated in a study that looked at burrowing owls in a New Mexico desert for sixteen years, ending in 2013 (note that the owl is found in the Mojave Desert, where the results would likely be similar, given the common desert environments). A predator that occasionally includes songbirds in its diet, burrowing owls (which prey more commonly on insects and small mammals) almost completely disappeared from the researchers' monitoring site. Where the researchers began the monitoring with fifty-two breeding pairs, they concluded with just one. The reason for this near-complete collapse in their population was, again, declining precipitation.[43]

Worse, non-native invasive annual grasses are exacerbating the damage by exponentially increasing the risk of damaging wildfire. Most of the southwestern deserts have such a grass; red brome is the scourge of the Mojave. These grasses close the gaps between the native plants, which enables a fire to spread from cluster to cluster of Joshua tree, creosote bush, or other native plant. Without the red brome, the bare ground between desert plants kept fires from burning more than a few plants. This means fire was an uncommon event in the southwestern deserts, so few plants evolved defenses against the primal force. Plants like the Joshua tree and creosote bush are easily killed by fire, and they don't have cones or seeds that survive the flames, as some western trees do. The invasive grasses cure early in spring, are flashier fuels for fire, and are worse in years of high rainfall.[44] There are far more fires now than there were historically, both because we're starting more (by means such as cigarettes and escaped campfires) and they're getting bigger vis-à-vis the invasive grasses. These two factors have brought about a million-fold increase in fire in the Mojave.[45] Once burned, desert plants do not quickly reestablish themselves, leaving the door open for the invasive grasses to take over the burned area.

Moreover, if business continues as usual, red brome and the other invasive grasses will find an environment increasingly suitable for them. Warmer winters will be more tolerable for the grasses that can't handle much frost, while longer summers will lengthen the fire season.[46] Rising temperatures and falling precipitation will make conditions intolerable for Joshua trees in 90 percent of their range, including most or all of their namesake national park.[47] With droughts worse than anything seen in the last millennium on tap if we don't curtail our emissions significantly and soon, the future of the Mojave and its iconic tree is bleak.[48] Indeed, if an experience I had in the southeast Oregon part of the Great Basin (the coldest of the country's five deserts) in 2012 is any indication, repeated fires could turn parts of the Mojave into biological deserts dominated by a single invasive grass. Cheatgrass, a cousin of red brome, infests much of the Great Basin, and just as red brome does, it closes the gaps between native plants (there dominated by sagebrush), thereby endangering an ecosystem in which most plants are killed by fire and few reseed. I drove through an area that had likely burned twice in recent years; the first fire killed the sage and other natives but left the thicker, live stems, which were consumed by the second fire, along with all the cheatgrass. The result was a lunar landscape bereft of anything organic, for mile after mile, as far as the eye could see. Not a living thing, stretching to distant hills on either side of the highway. The seeds of the next generation of cheatgrass were out there to be sure, but otherwise it was naked, charred land. Is that the future of the Mojave too? There are more parallels than differences between the two deserts, and with the Mojave already deeply compromised, we may well be approaching a tipping point beyond which the Mojave will not recover.

Some fifty miles before the end of the river trip, we rounded a bend in the canyon to find a deep, chesty, aquatic roar greeting us, along with a chunk of lava the size of a small house in the middle of the river. These were the signs that we were approaching Lava Falls, the most challenging rapid in the canyon at some river levels. There, the river drops thirty-seven feet—over a horizontal distance of just six hundred feet—producing the thunderous roar. We stopped above it to scout the best line through it, which is marked by a line of bubbles. How anyone can distinguish that line from the aquatic maelstrom is beyond my ken, but we successfully ran it with nothing but an oar going overboard (it was promptly retrieved). I don't know if I have ever held on for dear life as strongly as I did in Lava Falls, and to celebrate our success, our guide Ruth Ann broke out a beer for the four of us in her raft. Even though it was warm, Rainier never tasted better!

The western part of the canyon has many outcrops of basalt lava, all of it extruded in the last two million years from vents on the North Rim's Uinkaret volcanic field. Some of the lava eruptions cascaded down into the canyon, producing much taller lava falls that would have been spectacular to behold, a red-and-black seething mass of molten rock, slipping and falling thousands of feet down over the sedimentary layers. Today, fragments of lava stuck to the canyon walls tell us that, beginning about 725,000 years ago, the lava flows reached the river, damming it at least a dozen times, sometimes to depths of over a thousand feet. The higher dams backed the water up over a hundred miles, creating large, elongated lakes. In at least one instance, so much lava flowed into the inner gorge that, once the river was stopped, the lava flowed downriver eighty-four miles, substituting a river of molten lava for the water. Some of the reservoirs took years to fill and lasted up to twenty thousand years, but others lasted only till the river overtopped the dam and began eroding it. In some cases, the dam collapsed quickly, producing catastrophic floods as the large reservoir drained in a matter of hours. Then another lava flow would impound the river, and the cycle of creation and destruction would repeat itself, sometimes with a slow erosion of the dam, other times faster. The last one hundred thousand years have been quiet, allowing the river time to work at lowering today's Lava Falls. The falls may make for exciting rafting, but it's just a hint of what happened here in the not-so-distant past.[49]

Back in the rafts after the warm beer, we were essentially home free, with no major rapids between the falls and the takeout. We passed two more lovely days in the lower canyon, listening to the music of the canyon wren and happily ensconced within the desert Valhalla. After two weeks of river journey, we felt the current begin to slow as we approached Lake Mead. We enjoyed one more meal on the river's edge, this one featuring salmon and fresh asparagus, both brought to us by the captain of the jet boat that arrived soon after we'd made camp. The boat was our transportation back to the other world we inhabit, that of cities and highways, showers and comfortable beds, noise and smog, families and loved ones. As if to make us anticipate those hot showers even more than we already were, nature gave us a windstorm that night, coating our skin's layers of grit and silt with a layer of red sand icing. Perhaps the wind was nature's way of expressing her opinion about the reservoir impounding part of the grandest canyon on earth, a sentiment many of us would agree with. Whatever the case, we awoke to a hasty breakfast, and a fast trip out to Pearce Ferry, a marina some thirty miles across the lake.

The jet boat ride concluded the river trip in more ways than one, for it finalized some of the themes of this story, if in unexpected ways. Before we broke out onto the open lake, we traveled for miles below thirty-foot, flat-topped banks of silt on both sides of the river, which appeared there to be an arm of the lake (there was no current and the river was wider than it was in the canyon). The banks were actually deposits of silt and mud, from when the lake had been higher and the river dropped its sediment burden as it slowed and entered the lake. Mile after mile, for over ten miles, we passed this mute testimony of how much sediment the river carries—and that was just the accumulation of thirty years, the time between the completion of Hoover Dam / Lake Mead (1936) and the Glen Canyon Dam / Lake Powell (1966), which has been trapping the sediment since then. Like its neighbor upriver, Lake Mead has been dropping steadily since 1999, and was also about 145 feet below full pool when we boated across it. The decline in lake level has allowed the river to cut down through the sediments, transporting them downstream, closer to Hoover Dam. The banks of sediment were a testament to the Colorado River's sediment burden and erosive power.

The jet boat ride was also a foreboding illustration of the impacts of unrestrained global warming on us, for the decline in the lake level was obvious. If the projections are correct, the lake will never fill again (much like Lake Powell). The reservoir is one of the largest in the world, storing two years of Colorado River flow. As mentioned earlier, over thirty million people depend on the Colorado for their drinking water, and farmers irrigate five or six million acres of crops with its water. The river is so heavily used that little of it reaches the Gulf of California; instead, it comes out of the faucet in Los Angeles, Phoenix, and Las Vegas, brought to these cities by some of the longest and largest aqueducts in the country. The states through which the river flows have repeatedly litigated and negotiated with each other over its usage. Those negotiations are supposed to include consideration of water for Mexico, but the original allocations of water were based on very wet years. Thus, even before recent droughts, typical flows did not include enough water for Mexico. Further, the Southwest is one of the country's fastest-growing regions, so more squabbles over its water are inevitable, even without climate change. Add to this evaporation, which takes 3 percent of the water stored in Lake Mead; hotter temperatures will only take more, further compounding the tensions. Global warming and its impacts on people are here, now, just as they are for the piñon pine—for example, in 2018, the lake level fell to within two feet of the trigger for mandatory cutbacks. It's highly likely to fall below that level by 2026. The future climate of the Southwest is hotter and drier than it already is, and we have only begun to feel its self-inflicted impacts.[50]

With a final look back at the Grand Wash Cliffs, which mark the end of the Grand Canyon, we were soon at the marina. From there, a van whisked us back to Flagstaff and our vehicles. We gathered for one more meal together in the hotel restaurant that evening, and then went our separate ways the next day. Today, when I think back on the trip, I don't think of global warming or ALS (though they're never far from mind). I think instead of the beauty we experienced: two full weeks immersed in the most stupendous desert canyon on the planet. I think of the river, that passageway through the world of rock and cliff. I think of the life we encountered down there, the bighorn sheep, the lizards, the agaves. I hear the sounds of the canyon, the canyon wrens, and the rushing river—always the sound of moving water. I feel the desert warmth, the aridity, the refreshing dips at day's end. I sense the togetherness of backcountry travelers, the awe of that landscape, the timelessness of the journey through wilderness. Most of all, I feel the serenity and power of that place of nature.

Chapter Three

GLACIER NATIONAL PARK

Glacier National Park in Montana is all about its namesakes—past and present—and shining mountains. Glaciers past carved, gouged, and rototilled the mountains, while glaciers present, some two dozen of them, cling to life in high-mountain fastnesses. Mountains reach for the sky, their stratified layers telling us that, in the distant past, immense tectonic forces cracked the earth's crust and drove a massive slab of it over itself. Together, these forces of creation and destruction produced the vertical, yet lake-studded landscape we know today. Over 700 lakes and ponds can be found there, from tiny tarns that bear no name (only 181 are named) to ten-mile-long Lake McDonald. Situated on the Continental Divide, the waters from those lakes flow not to two oceans, but to three, the park being so far north that some of its water flows to Hudson Bay, part of the Arctic Ocean (which means there is a point in the park where the water splits three ways; fittingly, it's called Triple Divide Peak). The glaciers left deep valleys and high mountains, with over seven thousand feet of elevation between the extremes. Forests cloak the valleys, with the Continental Divide separating drier, sometimes scrubby forests on the east side from the more lush, sometimes ancient forests (dating to as old as 1517) on the west side, whose climate is somewhat moderated by the Pacific Ocean. The variable climates and elevations make for an exceptional biodiversity for such an inland, northern location: more than a thousand species of vascular plants call Glacier home, including thirty plants endemic to the northern Rockies and two carnivorous plants: mountain bladderwort and a species of sundew. Glacier is indeed a world-class treasure, an overload for the senses and a place of endless fascination for the student in all of us.[1]

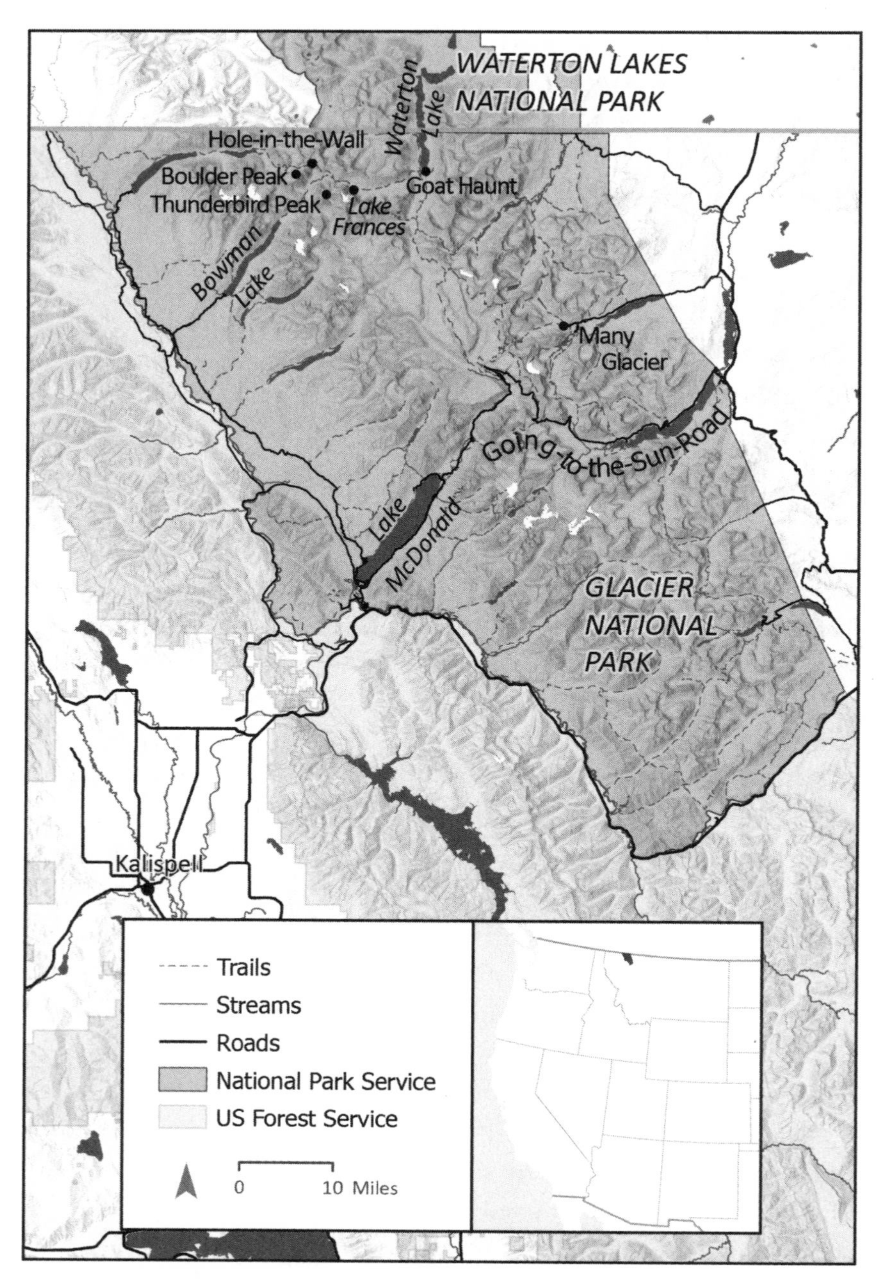

MAP 3 Map of Glacier National Park. *Sources*: NPS, USGS, Natural Earth, Natural Resources Canada

I first visited Glacier in 1991, early in my career with the NPS. Despite waking up one morning to three inches of wet, sloppy snow weighting my tent down, I saw enough of the park to understand that it was well worth the six-hour drive from Yellowstone. I became a regular visitor, chalking up sixteen trips there by the time of my diagnosis. Seeing the park one last time was high on the list of things I wanted to do in 2014, before my abilities became too compromised. At first, I planned to take a weeklong hike through some of Glacier's remote backcountry with some friends, but as the disease began affecting more vital activities like walking and balance, the plans shifted to less ambitious hikes, and finally, to sedentary pursuits like camping and auto touring. Glacier's magnificence might be more readily sensed in the backcountry, away from the roads and hotels, but it's pretty impressive from the front country as well, which is, after all, how most people experience it. In fact, the park's automotive centerpiece, the Going-to-the-Sun Road over Logan Pass, is one of the finest sightseeing and engineering accomplishments of any of the country's scenic roads. Practically glued onto the mountainsides in places, the road virtually demands that you marvel at the scenery—at least if you're a passenger in the vehicle, for the road's cliff-hanging nature requires the driver's close attention. America has many roads built to make fantastic scenery accessible, but few practically meld with the landscape to the extent that Glacier's road does, immersing you in its mountain paradise, wetting you with cold meltwater, and dropping your jaw in amazement around every turn.

Whatever the activities, I would be driving to Glacier not from Yosemite, where I had been working for four years, but rather from Yellowstone, where I had spent the first twenty years of my career. I owned a house in Gardiner, Montana, at Yellowstone's north entrance, and had lived in it for less than a year before leaving Yellowstone to work in Yosemite. I had bonded with Yellowstone and had always intended to return there, if for nothing other than retirement. Once I had the diagnosis, that return came sooner than expected, and under quite different circumstances. I would end up spending the summer and fall of 2014 living in my Gardiner house and working a reduced schedule in an office at Mammoth Hot Springs, the park's headquarters five miles away. To help me around the house, I invited a guy named Pete Whalen to live in my spare bedroom; in exchange for reduced rent, he would take care of the lawn and most of the cleaning inside.[2] For the first half of the summer, this arrangement worked out well (more on the second half of the summer later). Additionally, friends in Gardiner helped out with meals, especially Sean and Missy Miculka, two of my closest friends, who lived just a block away with their two young children.

A housemate wasn't the only indication that I was gradually losing my independence and my ability to live on my own. I continued falling, with two spills in one day when I was showing a visiting friend around the park. I was still using trekking poles for balance, but there was only so much they could do to keep me upright on the uneven terrain in Yellowstone. A week after the two falls, I took a two-mile walk with the Miculka family through the geyser basin at Old Faithful, and by the end of our walk, I was feeling shaky and uncertain of my balance. Although I didn't fall, it was clear to me that I wouldn't be doing any hiking that summer, which was heartbreaking to me. I had been looking forward to revisiting many of my former haunts that summer, given the abundance of two-to-six-mile hikes in Yellowstone, but that would not be happening. I had a lot of strength in my legs though, and a few of the trails I wanted to do were open to biking—so if I could find a recumbent tricycle (to provide the balance I no longer had on a standard bike), I could still experience those places. As it turned out, there was a firm in Salt Lake City that made such bikes, so Sean and I made a weekend trip there and came home with a trike that fit in the back of my Prius. I was set to go biking, and had fun going to some of my favorite places.

In early July, I flew to California for the next ALS clinic, at which a variety of specialists, including speech pathologists, physical and occupational therapists, pulmonologists, and neurologists, come together to see people living with the disease. From the beginning of that trip, it was clear that things were changing for me, and a lot faster than I wanted. Leaning on my wheeled suitcase at the San Jose airport, I fell backward onto the sidewalk, landing flat on my back. I was not injured, but the fall highlighted my decreasing health. Then, at the clinic, the speech pathologist noticed more decline in my legibility, especially with the hard "g" and "k" sounds (such as in the word "Greek"). She gave me some pointers on substituting sounds that would help the listener understand me. Most concerning of all was the recommendation from the physical and occupational therapists, who, as soon as they learned I had a history of falling, told me I needed a motorized wheelchair. I pushed back, asking them why they went straight to a wheelchair when a walker would suffice. They replied that wheelchairs like the one I would need usually take several months to get approved by insurers and then built. They finally agreed that a walker would help in the interim, but that I should start the process of obtaining a wheelchair. Not since the diagnosis had I gotten news this unwelcome. It was ugly proof that the disease was moving more rapidly than I had expected—and that I would soon need more care than a housemate could provide. I left the clinic and the state depressed and discouraged. This would not be the first time I felt like that way after an ALS clinic; as I was beginning to learn, there is never any good news with ALS.

Back home in Montana, I bought a couple of walkers (one for the car, one for the house) and made final plans for the Glacier trip. In late July, my friend Janet Hesselbarth picked me up in Gardiner, and we drove to the Many Glacier area on Glacier's east side, where we met up with two more friends, Charlie and Margaret Repath. I knew all of them from my work in the national parks, Janet from Yellowstone and the Repaths from Yosemite. Together we enjoyed a dinner of chili and chips and salsa, catching up and telling each other about our summer trips thus far. The next morning, we took off for the Waterton Townsite in Glacier's Canadian counterpart, Waterton Lakes National Park in Alberta. Our ultimate destination was Goat Haunt, which is back in Glacier at the head of Waterton Lake, a long lake that spans the international border (from the townsite to the haunt). In healthier times, I would have paddled the seven or eight miles, but we took the *International*, a passenger tour boat that operates out of the townsite four times daily, staying at the haunt for a half hour. None of us had been to Goat Haunt before, and in looking for new experiences that I could do in my compromised state, I had hit upon the idea of spending a few days there. With trails branching off from there, and with no road access, Goat Haunt promised plenty to do for my friends and natural tranquility to absorb for all of us. It would not disappoint.

Once the boat departed with its passengers, we had the small campsite to ourselves. Goat Haunt consists of the campsite, a ranger and US Customs station, a covered pavilion for the boat passengers to take shelter in during inclement weather, a bathroom with running water, and a few small houses for the employees stationed there in the summer months. Rising from the lake was a bevy of high mountains, with Olson and Campbell Mountains, both at least three thousand feet up from the lake, prominent in the view from the campsite. Down the lake were other peaks, pulling the eyes north into Canada. After we had checked out the surroundings and registered with Customs, we headed for the Peace Park Pavilion to cook supper. The pavilion commemorates the Waterton-Glacier International Peace Park, established in 1932 to celebrate the longest unguarded international border in the world. Both parks were established and are managed independently of each other, but they do cooperate with each other—for instance, working together to fight wildfires near the border (like the 2018 fire that threatened Goat Haunt as I wrote this very paragraph; it stopped well shy of the haunt). More importantly, the symbolism remains relevant today, reminding us of the value of peace. That is certainly what we found at Goat Haunt, and the pavilion became our place to hang out between the four daily boat arrivals.

After the last boat had left, a thunderstorm moved in. The pavilion was the perfect place to watch the storm, providing shelter from the rain but otherwise allowing us to feel the storm's power. Lightning danced across the mountaintops, illuminating them in momentary flashes of white hot light. First Campbell Mountain, then Olson, then a mountain down the lake would be flooded with a light not otherwise found in nature. Seconds later, thunder boomed, echoing back and forth between the mountainsides so loudly it seemed it wanted to knock the mountains down. Then came another volley of sky-rending sound, walking over the previous thunderclap's echoes, as if trying to outperform the earlier clap.

Rain came down in sheets, wetting our feet and shins as droplets blew in on gusts of wind. The same wind churned the darkened lake into whitecaps, breaking on the shore in front of us. It was nature's symphony, with the entire valley serving as its venue—the percussion section deserved a standing ovation when the concert ended a half hour later. Gradually, the light show moved down the valley and the echoes subsided in intensity and tempo. The day could hardly have ended more powerfully, more memorably, more beautifully.

The next day, Margaret and Janet took a hike while Charlie stayed in camp with me, enjoying the quiet and exploring the immediate area, where the walker could go. Lake Janet was the women's destination, an understandable one for our companion, but for more than the obvious reason. Seven years earlier, her late husband, Woody, and I had hiked to Lake Frances, the next lake up the valley, some two or three miles upstream from Janet's namesake. She would not have enough time to get to Lake Frances, but being in the vicinity would allow her to touch his spirit. The visit that Woody and I paid to Lake Frances was part of a longer hike that brought us to some of Glacier's wildest and most vertical scenery, via some of the park's most impressive trail building. The same hike brought us face-to-face with some of climate change's effects on Glacier, both obvious and subtle. For that reason, let's go back to Glacier's Hole in the Wall with Woody and me, back to a hike that was as beautiful as it was sobering.

In early August 2007, we began our trip not by putting our boots on but by launching my canoe into Bowman Lake, in Glacier's northwest corner. Bowman is one of eight long and narrow lakes (including Waterton Lake) that emanate from the park's high-mountain core. These lakes all owe their existence to glaciers past, rivers of ice miles long that ground and scraped their way down from the park's highest peaks to its lowlands. Along the way, they broadened and deepened the valleys and carried the sand, gravel, dirt, and boulders that they exhumed to where they melted, leaving them in a pile called a moraine. Once the glaciers had melted off, those moraines became dams creating the lakes that are such a part of the Glacier experience today; almost all of the park's popular tourist destinations have lakes in the foreground. For Woody and me, Bowman Lake was a seven-mile-long aquatic portal into Glacier's wild high country. Also, since we didn't have to carry our backpacks that distance, we were able to shave a day off of our approach at both ends of the trip.

The northern part of Glacier is in the rain shadow of the Whitefish Range, which wrings some of the moisture from the storms blowing in along the prevailing storm track. By the time clouds rise up and over the Whitefish Range, dropping a load of snow and rain there, they don't have as much moisture to give up when they collide with the mountains in northern Glacier. For this reason, the forests that Woody and I paddled past were dominated by lodgepole pine, which prefers drier habitats than the cedars and hemlocks found around Lake McDonald. Without a mountain range intercepting Pacific storm systems, Lake McDonald enjoys a wetter climate. For the same reason, the glacier that created Lake McDonald may have been larger, carving out a deeper hole in front of its moraine and producing the deepest lake in the park. At 464 feet deep, Lake McDonald is so deep it sometimes does not freeze over completely in winter. (Cold water is heavier than warm water, so as air temperatures fall, the cold surface water sinks, bringing up warmer water from below. Not till the entire lake is cooled can the surface freeze, which means deep lakes take a long time to freeze, if they do at all.) Bowman Lake is 253 feet deep, so it takes a few months to freeze, but it usually does fully freeze over. It's remarkable how Glacier's landscape is a reflection of its glacial past, even whether the lakes freeze and the general date of freeze-over. We would continue to see the glacial influence, past and present, as we paddled onward into the wild.[3]

A couple of hours of paddling brought us to the head of Bowman Lake. The easy traveling was over, so we hid the canoe in a thicket of trees, shouldered our backpacks, and bushwhacked through the trees to the trail, which parallels the lakeshore. For the next four miles, the hiking was fairly flat, proceeding up the Bowman Creek Valley. Then, the trail crossed the creek and gradually began climbing away from it, eventually turning east into a tributary valley. By this point, lodgepole pine began to disappear, replaced by subalpine fir and Engelmann spruce, two trees that are better adapted to heavy snow. Crossing the tributary stream, we were soon switchbacking up the far side and breaking out of the forest cloaking the Bowman Creek Valley. The sides of that small valley are too steep for snow to cling to, so avalanches are common. Some slide so regularly that they have prevented trees from establishing in their slide paths, while others occur so infrequently that trees take root, only to be taken out when the avalanche reclaims its path, snapping the trees off like they were matchsticks. The end result is the same: an avalanche scar that, in summer, becomes a meadow that allows—no, the proper verb is "demands"—that you pause to look up and around. The slopes above us and across from us soared four thousand feet and higher into the sky, their summits out of our sight. Meanwhile, at our level grew a profusion of wildflowers and shrubs, all of them short enough to be protected by the snowpack when avalanches come roaring down from the heights above. They, too, are nourished by the snow that melts and seeps into the soil, sustaining a higher soil moisture than rain alone would create. We were surrounded by signs that snow, not just in the form of glaciers, is crucial in creating and sustaining the Glacier landscape.

It's hard to overstate the importance of snow to western ecosystems. In contrast to eastern and prairie ecosystems, which receive much of their annual precipitation in the spring and summer growing season, the eleven western states receive the majority of their annual precipitation in the form of snow. Come spring, the snow begins to melt, percolating into the soil and recharging the groundwater. The soil may dry out on the surface, but the groundwater is usually sufficient to carry trees through the rest of the growing season. Summer thunderstorms supplement the snowmelt, but their influence is minor compared to snowfall, especially in the three Pacific coast states, dominated as they are by a Mediterranean climate of wet winters and bone-dry summers. The Four Corners states (Arizona, Colorado, New Mexico, and Utah), where the summer monsoon brings wetting thunderstorms, are a partial exception to this generalization, but even there, the dominant trees still depend more on snowmelt than on summer thunderstorms. The northern Rockies (which include Glacier and Yellowstone) fall between these two groupings, receiving more summer rainfall than the coastal states, but less than the southwestern states. With more than 80 percent of annual precipitation falling as snow, northern Rockies ecosystems depend more on snowmelt than on summer thunderstorms. In the past, by the time snowmelt and the associated groundwater began to taper off, the summer growing season was drawing to a close, so there was only a short window of suitable conditions for wildfires to burn, at least in most years in the northern Rockies. As we'll see soon enough, anthropogenic climate change is altering that relationship, with predictable and fiery consequences.

Before long on our hike, the trail began to level out, telling us that we were approaching Brown Pass. We left avalanche scars behind, entering a flat area with a lovely blend of clumps of small trees distributed polka-dot style on flower-strewn subalpine meadows. We passed scattered stalks of bear grass, its three-foot stalks of creamy white flowers (the symbol of Glacier to me) gone to seed. In full bloom, though, were vibrant clusters of fireweed, its pink and purple flowers the harbinger of autumn—in early August—for fireweed is generally the last flower to bloom in the northern Rockies. Brown Pass is just below treeline, which is about seven thousand feet in Glacier, so any plants there are exposed to hurricane-force, arctic winds in winter, winds that scour and abrade anything that sticks out above the snow. For these reasons, trees are often stunted at or near treeline, with their upper branches appearing threadbare and pointing away from the prevailing wind direction. Others take on a low-profile, spreading and crawling form known as krummholz, a form that traps snow between the knee-high branches, thereby giving the low-slung trees protection against the scouring, frigid winds. Given the harsh conditions there and above treeline—frost and snow can occur any month of summer—plants grow extremely slowly there. Three inches of new growth is exceptional for a tree at Brown Pass, and trees just two inches in diameter at the four-foot-high level can be hundreds of years old. The subalpine beauty through which we walked at the pass was still another example of the influence of snow and winter on Glacier's landscape, one that is as cold as it is beautiful.

At the pass, we reached a trail junction, one that would take us to the Hole in the Wall. We had hiked a horizontal seven miles and climbed a vertical half mile, and had just two horizontal miles to go before we reached camp for the night. We turned left and began ascending a gentle slope covered with wildflowers and krummholz that gave us incredible views of Glacier's high, layered mountains and snowfields. Glacier's mountains are composed of numerous layers of sedimentary rock, much like those of the Grand Canyon, except that Glacier's layers are both more numerous and much thinner than the Grand Canyon's. Also, the layers of both parks were exposed by uplift and erosion, but Glacier's layers have the added step of being thrust over much younger rock layers, as mentioned above. About 150 million years ago, North America collided with another plate of the earth's crust, resulting in immense compression forces in the West that led to the uplift of the Rocky Mountains. In the Glacier area, those forces were so strong that the crust buckled and tore, with a wedge several miles thick and hundreds of miles long riding up, over, and onto itself—and then sliding some fifty miles east as the compression forces continued. The action stopped about 60 million years ago, and erosion and glaciation have been at work since, leaving us with the layered mountains we so love today.[4]

On our hike, those layers soon became more evident as the trail began leveling out while the slope steepened. The slope steepened further, becoming near vertical, while the trail stayed level, following one of the layers across a mile-long cliff face. In some places, the exposed layer was just three feet wide, so the trail used the entire ledge. There were no cables to hold on to, so we had to rely only on our senses and judgment, though I don't remember anything too marginal. We couldn't walk and admire the scenery at the same time, for one misstep could have sent us falling down the mountainside four to seven hundred feet. So, we stopped frequently, amazed by the colossal views. Across from us rose the Sentinel Peak and Thunderbird Mountain, one more pyramidal than the other and both flecked with snowfields and a small glacier that together spawn Thunderbird Creek. Partway down from the pyramids, the stream gathered its waters and jumped out over a ledge into space, dropping a hundred feet before disappearing behind an intervening ridge. To the right of the pyramids, Boulder Peak shot up four thousand feet from the Bowman Creek Valley (see plate 12). Only the lowest thousand was clothed with tree or shrub; the rest was crag or cliff, pinnacle or spire, loftier and loftier. This was Glacier at its best, rock layers that made for vertical and horizontal beauty and excitement.

Shortly after gazing upon Boulder Peak, Hole in the Wall appeared around a bend in the mountainside we were hiking across. Named for a spring and waterfall that emanate from a small cave, the term now seems to apply to the entire subalpine basin we'd be camping in. Hole in the Wall is a bowl about a half mile in diameter and 6,400 feet high, with a sprinkling of trees similar to what we had found at Brown Pass. The bowl is open on the downstream side, the valley floor dropping away 1,500 feet to the larger valley below. Hole in the Wall was once occupied by a glacier, one that joined a larger glacier occupying the larger valley below. The larger one had more scouring power, carving the lower valley much deeper than the small tributary could do to Hole in the Wall. The result is a hanging valley, a side drainage that ends much higher than the main valley and whose stream often goes over a tall waterfall or cascade where the hanging valley ends. Hole in the Wall actually has two falls, the nascent Bowman Creek and Hole in the Wall Falls, the latter emerging from a cave. It was still another example of why this is *Glacier* National Park; indeed, we would see more hanging valleys—and more evidence of glaciers, past *and* present—as we continued our journey.

But first, we set up camp for the night, after sixteen miles of wilderness travel. Our home for the next two nights was the Hole in the Wall backcountry campground, located on a knoll at the back of the hanging valley. Like all of the park's wilderness campgrounds, this one accommodated several hiking parties, with three to six tent pads scattered around a common food preparation area and food storage pole (to hang the food out of reach of bears overnight). This one had a view better than most: framed between the sides of the hanging valley was the Sentinel Peak / Thunderbird Mountain massif, the two pyramids pointing to the heavens. It was a magnificent view—but not one, evidently, to enjoy from our campsite, which was tucked into a copse of trees. Someone had worked long and hard to find a grove of trees, which are hard to come by there, big enough to completely conceal a half dozen tents. What's more, they had worked harder to ensure that the tent pads stay within the trees by hemming the pads with logs and other chunks of wood. Unfortunately, their zeal to protect nature meant that we had nowhere to go to enjoy the view we'd worked so hard to enjoy. It was as if the guide leading you through the Sistine Chapel made you wear a blindfold while inside the building but then let you look through the windows from outside. Woody and I understood that to protect nature means you have to tolerate some human impact, for few people would support the preservation of a special place in nature if they could not enjoy it in a reasonable manner. For that reason, we cleared a small spot in the dead wood to allow us to sit and watch the shadows lengthen. The ground we uncovered was not vegetated, indicating others had done the same thing—and suggesting the futility of prohibiting the faithful from seeing that which they had labored to experience, within commonsense limits. We left our viewpoint and hangout cleared off, for future campers to enjoy.

The next day was a layover day, a day for us to explore the local neighborhood. We took an eight-mile day hike, to the summit of Boulder Peak and back. I remember the day clearly; it was one of those rare days when things couldn't really be any better, a day bordering on perfection, a day where the senses are in overdrive much of the time. Our hike took us around our hanging valley on to a flat bench, then through Boulder Pass to the northwest ridge of the mountain, which we climbed up to the top. The hike was mostly above treeline, and the skies were clear, giving us world-class views every step of the way. Mountains piled upon mountains clustered together all around us, near and far, and we were just after the peak of the alpine flower bloom, so there was a lot of color along the way. Magenta Lewis's monkeyflower lined a meltwater stream, while lavender asters, red Indian paintbrush, and yellow daisies brightened patches of meadow. The worries and cares of the world dropped away, enabling us to forget that Woody was facing some serious health issues (more on this later). He was always a lighthearted fellow, but that day his joy was overflowing. There was beauty and wonder on many scales to be sensed, from the stupendous mountains to clusters of wild onions poking through moss on the edge of an alpine pond; from the stalwartness of a tree seemingly growing out of solid rock (the tree's branches hid a large crack in the boulder) to the audacity of tadpoles at the edge of another pond, cold-blooded survivors in a place that's snow covered eight months a year; and from the headiness of peering out over a four-thousand-foot glacially carved abyss to the tranquility of sunset over the two pyramids, both of them catching the sun's last rays. For these reasons and more, that day has seared itself into my basic conception of Glacier, for my dreams of the place today are filled with color, sensory overload, and peace.

As if all that were not enough for one day, the landscape also had some geologic features that further enriched the experience, if one recognized and understood the clues. One of these was a pattern of ripple marks on one of Boulder Peak's namesake rocks. Just as you see today in shallow water, ripple marks can be preserved if the substrate in which they occur is transformed into rock. They add a dimension of texture and beauty to the rock, and can be used to determine which side was up when the marks were formed: the crest of the ripples on this boulder was up. Another find was a pair of concentric half circles in a rock outcrop the trail crossed; they measured six to eight feet in diameter, each with some two dozen banded semicircular layers. They were impressive enough to photograph, though not till I did the research for this book did I realize that they were stromatolites. These are the remains of the earth's oldest life forms, cyanobacteria (or blue-green algae), which are 1.6 billion years old in Glacier, but up to 3.7 billion years old in Greenland. They grew in colonies, binding and cementing fine grains of sand and other sediments into place as they grew, eventually forming layered domes (which means there are layers within Glacier's layers). The half circles we saw were cross sections of the colonies, which grew in shallow water. Stromatolites are cool not just due to their age, but also because they put oxygen into our atmosphere as they photosynthesized over billions of years. In other words, by putting what animals need to breathe into the atmosphere, they made it possible for humans to live on planet Earth.[5]

It's no surprise that evidence of glaciers was all around us that day, given all the evidence we had been seeing thus far. U-shaped valleys flowed away from us in all directions, rounded out by the rivers of ice that once occupied them. We passed another hanging valley whose glacier not only carved it out but also scraped all the soil away, leaving just naked rock, a few puddles of water, and a few stunted trees scrabbling for soil. We could see three glaciers from atop Boulder Peak: Thunderbird, Agassiz, and Weasel Collar. But perhaps the most revealing sign of glaciers was something we didn't see: Boulder Glacier, which once covered the north side of the mountain we were on, sliding all the way down to the pass a thousand feet below. Historic photos of the glacier, taken between 1910 and 1932, show it flowing down from near Boulder Peak's summit and across Boulder Pass (entirely filling it), bumping up against the far hillside, and then flowing a short ways toward Hole in the Wall. The terminus, easily twenty feet high, featured the opening of an ice cave tall enough to walk into, something visitors on pack trips to the glacier did until it disappeared. Probably already retreating in 1910, the glacier had melted out completely by 1998. Boulder Glacier is history, replaced by seasonal snowfields that usually don't last the summer.[6]

Shrinking and disappearing glaciers are the story all over Glacier National Park. When the park was established in 1910, there were as many as 150 glaciers, judging by their telltale moraines. In contrast to the large moraines (like the one that created Bowman Lake) left behind by the much larger glaciers from the last ice age (which ended about 12,000 years ago), these moraines are smaller—but still as much as two hundred feet tall—and generally up high, not far from their glacier's source. Those alpine glaciers first developed about 7,000 years ago, reaching their maximum around 1850, at the end of a 450-year globally cool period known as the Little Ice Age. When that cool period ended, the glaciers began retreating, losing 73 percent of their surface area by 2003. That retreat has accelerated more recently, leaving just 26 glaciers in 2015 (they aren't measured every year), all of which continue to retreat. They vary in the rate of retreat, but if those rates continue, most of them will be gone by 2030, and all of them by 2080. Disappearing glaciers, then, have been an increasing part of the Glacier story, as they will for decades to come.[7]

As expected, the glaciers are melting because the temperatures are rising, with increased summertime warmth being the biggest factor in the retreat.[8] By almost any measure, Glacier is getting warmer. While western

Montana has warmed 2.3 degrees since 1900, Glacier has warmed even more.[9] Summer temperatures have risen 3 degrees, with summertime frosts at or below 5,700 feet in elevation virtually gone.[10] Extremely hot days (above 90 degrees) have tripled in number and are occurring over a summer twenty-four days longer than it was a century ago, while in winter, extremely cold days (below 0 degrees) are declining in number and ending twenty days earlier.[11] Overnight low temperatures are increasing markedly throughout the year, but especially in winter: 6.8 degrees since 1983, with the annual average increase in low temperatures just half a degree less.[12] The increased winter overnight lows are at the extreme warm end of Glacier's historic weather, pushing outside of Glacier's norms.[13] The annual number of nights below freezing has dropped by fifty-three, with twenty-one of those nights occurring in spring and another eleven in fall. While some of this warming is due to hemispheric weather phenomena like El Niño and its alternate La Niña (at least 8 percent), the most they cause is just 41 percent.[14] Between 59 and 92 percent, in other words, bears humanity's signature—roughly a month of freezing nights gone, thanks to anthropogenic global warming. It is clearly a warming climate in Glacier.

Thanks to the warmer temperatures, it's also a drying climate. Just since 1969, the peak snowpack occurs two or three weeks earlier, from April 15–20 to April 1 or earlier, which means that the spring melt is beginning that much earlier. Also since 1969, the number of days with snow on the ground has declined by fourteen days, with eight of those occurring in spring.[15] Warmer and more variable winter and spring temperatures are bringing more precipitation in the form of rain, not snow.[16] Record summer and winter droughts have occurred in the 2010s—as have record and near-record snowpacks, but even those have not stopped the glacial retreat[17] (or helped drought-stressed trees, as the experience of 2018 illustrates, where record snowfall was followed by wildfires that closed the west side of the park for half the summer). The snowpack is declining as well (about 15 percent since 1970), so much that the late twentieth-century reductions across the northern Rockies are almost unprecedented in the last millennium, as seen in the tree-ring record.[18] As with the higher temperatures, some of the snowpack decline can be attributed to hemispheric climate cycles, but at least half, and as much as 80 percent, of the loss is anthropogenic.[19] The net effect of these changes is that winters are shortening, snowpacks are declining, and summers are lengthening, all of which give more time for forests to dry out and burn.

As discussed above, Glacier's glaciers are melting faster due primarily to warmer summers, though declining snowfall clearly contributes to the trend. As with each of these factors, the human contribution to the glacial melt has been quantified. While our contribution to the natural background rate of melting since 1850 is modest (25 percent, with a range of 0–60 percent), it jumps to 69 percent from 1990 to 2011 (a range of 45–93 percent).[20] The melting is enhanced by a positive feedback loop, which is a situation in which a primary action causes a change that then stimulates more of the primary action. In this case, when part of a glacier melts off, it exposes the surrounding boulders, dirt, and cliffs, all of which have less albedo than ice and snow, so the uncovered substrates absorb more solar radiation, which warms them and their surroundings, including the glacier. The glacier then melts more, exposing more heat-absorbing rocks, and so on.[21] Positive feedback loops vary in the extent to which they are effective, and this one is probably just effective locally. There may, however, be another that is effective more regionally. As the lower elevations in the Glacier area warm, they receive less snow, and melt out earlier in spring. The lack of snow on the ground means that it can absorb more solar warmth, heating the ground and the air above it. Breezes blow the warmed air to the higher elevations, where it expedites the snowpack melting. This process happens every spring, but global warming is enabling it to begin earlier now, furthering the snowpack melting. Moreover, the advent of wintertime rain, which may prevent a snowpack from establishing in lower elevations (or melt it out), may be enabling this positive feedback loop to operate in winter.[22] Anthropogenic warming is thus causing multiplying melting effects on Glacier's glaciers.

Looking to the future (and using the Northwest Climate Toolbox), the warming is expected to continue and intensify. An additional 8.5 degrees of warming is expected at the elevation of Logan Pass (6,647 feet) by the end of the century if we continue business as usual, for a combined total of 11 degrees. The heating will be fairly uniform across the seasons, but winter overnight low and summer daytime high temperatures will see the greatest changes. The average winter lows are expected to rise from 10.6 degrees historically to 22.9 degrees, with the summertime high increasing from 60.5 to 74 degrees. Both of these increases will certainly further the glacial melt-off, but two of the lower-magnitude changes may have a greater effect on the remaining ice. By the end of the century, the average wintertime daily high temperature will go above freezing for the first time in history, to 32.8 degrees (from 23.5), and the average springtime overnight low will rise to nearly freezing (31.7 degrees, from 22).[23] Averages, of course, mean that some temperatures will be lower and some higher; as expected, it's the higher ones that concern us here. Temperatures above freezing in winter mean that some precipitation will fall as rain, not snow. Small amounts of wintertime rain can be absorbed by the snowpack, but larger amounts will melt it. What snow survives the warmer winters will melt earlier as a result of the increasing spring temperatures, especially the warming nights. As long as overnight temperatures get cold enough (generally 28 degrees or below) to refreeze the snowpack, the reduction in snowpack is modest, even on warm days. The warmer overnight lows of springtime will not always refreeze the snowpack, expediting the melt-off.[24] The two weeks that melt-off has already advanced will become a month, then six weeks, then possibly two full months by century's end.[25] Glacier's high-elevation climate, in short, will be heating up dramatically, with predictable results for the namesakes that still hang on.

Glaciers aren't the only thing that will be melting as this century progresses. Ice patches, just like glaciers, form in places where more snow accumulates than will melt in a summer. Over time, the weight of new snow compresses older snow below it into ice, but unlike glaciers, ice patches will remain stationary. This means that anything left on the surface of an ice patch (such as windblown fragments of wood and animal scat) can become encased in the ice, remaining in place for hundreds, and perhaps thousands, of years. Prehistoric Native Americans frequented ice patches to hunt game driven into the high country by biting insects and summertime heat, so prehistoric artifacts may be frozen into the ice too. Layers in an ice patch can be dated in the same way as the Greenland ice cores; some of Glacier's ice patches are six thousand years old. By removing ice cores and collecting items as they melt out, researchers can open a window into Glacier's past. Thus far, Glacier's ice patches have revealed a complete bison skull, indicating that bison used the high country (imagine seeing a one-ton bull bison silhouetted against the skyline of an alpine ridge!), and a fragment of yew wood, indicating the presence of a warmer climate in the past (which is not a reason, again, to dismiss concerns about the current global warming, which is happening at a rate faster than many, if not most, plant species will be able to respond to). No human artifacts have turned up yet in Glacier, but that is likely to change with time and more melting. Ice patches and the information they contain, then, are another treasure threatened by climate change.[26]

On Boulder Peak that glorious August day, Woody and I stayed on the summit for over an hour, eating our lunches, soaking up the sunshine, and admiring the 360-degree panorama. Of the three glaciers we could see, Thunderbird Glacier will be the next to go—if it hasn't already—because it just barely met the minimum size to be considered a glacier in 2015, when it was 26.7 acres in size (the minimum is about 25 acres). Agassiz will probably go after Thunderbird, even though it is larger than Weasel Collar, because Agassiz has been retreating faster, having lost more than half its surface area since 1966.[27] Seeing glaciers in the distance was nice, but seeing them up close, as I did years later in Olympic, is quite another experience. Woody and I didn't have that experience in Glacier that day, which would have driven home the connection between glaciers past and the landscape today. It will be a sad day when this tangible connection to Glacier's past is gone completely. For the two of us, the day was a home run, but seeing Boulder Glacier would have made it a grand slam.

We left the summit reluctantly, retracing our steps down the ridge and back on the trail through Boulder Pass to Hole in the Wall. As we hiked, we went back below treeline, that line above which climatic conditions are too harsh for trees. There is evidence that treeline is rising in Glacier, as seen in comparisons of historic photos with contemporary ones. Some show trees establishing at higher elevations than they did previously, others show them getting taller, and still others show them filling in open areas between patches of trees in the historic photos. Ecological change is slow to happen in these harsh environments, so we don't often see it occur, and the historic photos were usually focusing on a nearby glacier or mountain, so we sometimes have to look at the photo margins. But it's there nonetheless: as Glacier warms up, trees are responding, moving into places that were once too cold. And as we would see soon enough on that hike that other trees—many others—are responding in a quite different manner.[28]

At particular risk are plants and animals dependent on a cold climate. Several animals depend on the cold, season-long water emanating from glaciers and ice patches. This water is not only cold, but it is also abundant in late summer, when up to a third of the water in some Glacier streams is glacial meltwater.[29] The western glacier stonefly and the meltwater lednian stonefly, for example, are both restricted to short stretches of cold, alpine streams, often just below the bodies of ice. Both stoneflies decline in abundance downstream of the ice, and both are candidates for Endangered Species Act (ESA) protection due to climate change–induced habitat loss.[30] The stoneflies might be occasionally eaten by bull trout, which are already a threatened species under the ESA, due to habitat loss throughout their range, including that caused by climate change. They require the coldest water of any fish native to the Rockies—58 degrees, and just 48 degrees for spawning—making them dependent on the same streams (and headwater lakes) as the stoneflies.[31] Another species at risk is the black swift, which only nests near, or even behind, waterfalls that persist through the summer—those fed by glacial meltwater, not just by rainfall. Glacier has an abundance of such falls now, giving it more than half of Montana's nesting population of the birds (nine nests in Glacier, though there are likely more).[32] Still another bird species dependent on consistent streamflows is the harlequin duck, a brightly colored duck that is the envy of whitewater rafters, floating and bobbing its head through just about any whitewater stream. The duck sometimes builds its nest near the whitewater, so if the stream floods higher than normal, the bird will lose its clutch. Climate change is expected to make streamflows more variable, which will challenge the duck's ability to sustain itself.[33] The future for all these species is uncertain at best, with glacier-fed streams projected to have the greatest amount of warming in the decades ahead, as their glaciers and ice patches disappear.[34] In these and other ways, the disappearance of glaciers has real effects on wildlife that depend on the cold meltwater that they provide. The loss of glaciers is more than just aesthetic.

Other plants and animals depend on the cooler temperatures found in the park today (relative to those at the end of this century). Of the thirty species of northern Rockies endemic plants found in Glacier, all but one is found in cold, open areas (such as above treeline), and many are relics of the time when the large glaciers had just melted out. They persist in the park because it has pockets of habitat similar to what these species found ten or twelve thousand years ago—pockets that may not survive the warming to come. Indeed, some alpine plants have already shown declines in distribution since 1988.[35] For example, an alpine variety of saxifrage has already significantly dropped in abundance due to the incursion of wildfire into Glacier's alpine tundra, a phenomenon seen only rarely before. Such wildfires will exacerbate the effects of continued warming, increasing the projected loss of habitat for the saxifrage from 38 percent to 43 percent over current levels by 2050—assuming we take steps to reduce emissions.[36]

Some of these plants are probably among those gathered by the pika, a five-inch-long member of the rabbit family that frequently lives near or above treeline. Pikas don't hibernate, so they spend the summer gathering grass and other plants, drying them in hay piles near their burrows. Built to conserve body heat and stay warm, they have small ears and thick fur, and they always live near talus piles (mounds or slopes of fragmented, plate-shaped slabs of rock with spaces between the rocks), where they can escape afternoon heat. They can't effectively vent excess body heat, so sustained temperatures above 78 degrees are deadly. Some may seek the cooler temperatures at higher elevations, but once the pika reaches the summit, this strategy fails. For this reason, pikas are dying out all over the West, making the adorable lagomorphs the poster child for climate change outside of the Arctic (where the polar bear is the unfortunate star). Many times, I have heard their distinctive alarm call on a mountain climb, and with up to 9,800 pikas found in Glacier, we may well have heard one of them going to or from Boulder Peak.[37] As with the other animals and plants mentioned above, the loss of the pika would hardly be noticed by most people, but it would be one more thread gone from the rich tapestry of life, the treasure that is our common heritage in Glacier National Park.

Woody and I spent another tranquil evening watching the sun set over the pyramids, then broke camp the next morning. We began the day's journey by hiking back to Brown Pass via the ledges in the near-vertical mountainside, which was again a riot of fantastic views. We saw Bowman Lake far below, a sinuous aquatic blue between thickly wooded ridges. It would be our destination in twenty-four hours, but we had a different lake in mind for this day: Lake Frances, which is the other direction, down from Brown Pass. But first, we dropped our packs at the Brown Pass campsite and set up the tent. I remember that campsite mainly for its lack of water. The nearest source was a mile away, which we'd discovered when we passed through there two days before. We needed water to cook our evening meal, so rather than toting it a mile, we would bring the dinner and cooking supplies with us to the lake, which would have a much better view than the Brown Pass campsite. With the tent set up, we took off on the afternoon side trip to Lake Frances.

From the pass, the trail dropped 500 feet down a lush, forested hillside, the spruces and firs sheltering a rich understory of bear grass, huckleberry, and other shrubs. The forest was open, giving us views of Thunderbird Falls in its entirety, a stairstepping liquid lace cascade swallowed by shrubbery after 1,000 feet of air. In the distance, Mount Cleveland loomed, looking presidential, every bit the tallest, most respectable mountain in the park, at 10,479 feet above sea level. At the bottom of the hill, the trail emerged from forest to Thunderbird Pond, its water slightly cloudy with glacial flour (fine-grained silt) brought down by the creek from Thunderbird Glacier high above. On its shore was a rock big and flat enough to serve as our kitchen and dining room, so we decided to eat there, after we went to the lake and back. Another mile down the valley was Lake Frances, which was like a grown-up version of Thunderbird Pond—or the pond on steroids. For starters, it was a larger body of water, befitting the larger descriptor. It too had a meltwater stream tumbling down from the lofty heights, but this cascade was at least 1,500 feet tall. The lake was also colored with glacial flour, courtesy of Dixon Glacier, which is a little bigger than Thunderbird Glacier (they're in a losing race to hang on to existence). Finally, the lake was closer to the mountain rising above it, a shoulder of Thunderbird Mountain, which rose 4,000 feet above the lake. The mountainside was a medley of green patches of steeply pitched meadow, bands of layered cliff, and vertical ravines. It was, like many lakes in Glacier, a place where you felt small, no more than a speck of life.

The lake had a small beach, and we were dirty from three days of wilderness travel, so it was time for a swim. With no one else around, we swam in the buff like many wilderness travelers do. The lake was predictably cold, its water having been part of a glacier not long before, but it was not intolerable, especially on a warm afternoon. One of us found a boulder a hundred yards out, barely submerged, with only a few inches of water covering it. It was a perfect swimming destination, and it was easy to climb onto. I took a picture of Woody on the rock pantomiming a frontier explorer and pointing to the glacier far above, his pearly-white butt shining brightly for all creatures to see. It was the perfect representation of his free spirit, and something I will always remember him by. Woody's oncologist had called right before this trip to tell him she'd found active carcinoids in his lungs again. Discussing the situation with him, she encouraged him to do the trip with me and worry about it later. After the trip, he spent much of the next few months going through chemotherapy and radiation, but to no avail. Less than a year after this trip, he died of a rare form of lung cancer, at age fifty-nine. I think he knew his time was drawing near, for he seemed especially happy on the trip, even for him. Maybe that's why he was beaming in every photo I took of him—especially on the pantomime rock.

While we were swimming, some smoke blew in from fires burning west of the park, enveloping the mountains in a light-gray haze. We'd been flirting with smoke on the whole trip, having driven through it to get to Bowman Lake, and then seeing it to the south when we were on Boulder Peak. Smoke and the fires that create it are an increasingly common part of summer in the West, burning through millions of acres every year—with the trend continuing to go up. This is the other way plants are responding to climate change: by dying and burning. Glacier is no exception to this trend, with the worst fire season in Glacier's history occurring in 2003. More than 135,000 acres burned in six large fires, starting in mid-July and burning actively into early September. Like most forest fires, these illustrated the pivotal role of moisture in fire ignition, behavior, and eventual termination. Moisture levels in dead wood were at record lows when thunderstorms on July 16 ignited several of the fires. The fires spread rapidly throughout the next eight weeks, when no significant precipitation fell, and they finally subsided in intensity on September 8, when heavy rain fell throughout the park. By summer's end, one out of every five acres of forest or meadow in Glacier had burned.[38]

Forest fires are nothing new in Glacier, with the second-largest acreage burning in 1910, the year the park was established. About a hundred thousand acres were affected by fire that year, most of it in the southern part of the park, though half the northern shore of Bowman Lake burned. Much of that area is now covered with lodgepole pine, with its serotinous cones. These specialized cones have resinous bonds inside that prevent the cone from opening until a fire melts them. The tree is often killed by the canopy fires that tend to burn in lodgepole forests, in part because the pines frequently grow closely, allowing the fire to jump from tree to tree. Fires in lodgepole forests don't occur often, generally only once a century or so, but when they do happen, the serotinous cones open, starting the next generation of trees. A different adaptation to fire is found in Douglas fir and western larch, two other common trees in Glacier's west side. These trees have thick bark, which insulates the growing tissues underneath (the cambium) from the heat of ground fires. This adaptation is good in places that burn frequently, as often as every ten or twenty years in the northern Rockies. These adaptations suggest that fire has been around long enough for plants to evolve to cope with it.

In the arid West, fire is not just a destructive force to contend with—it's an essential nutrient recycler. Decomposition takes place quite slowly in the West; there is too little moisture in summertime, when the organisms that break down plant materials are active, and in winter, when moisture is plentiful, temperatures are too cold for them. Consequently, fire takes the place of decomposition to various extents throughout the West, depending on local conditions. Recognizing this and other benefits of fire, the NPS has permitted some fires to burn under careful monitoring and suitable weather and fuel-moisture conditions (known as prescriptions), since the late 1960s. Under this "prescribed fire policy," lightning-caused fires have burned a modest number of acres in Glacier. Other fires, including all human-caused fires and those lightning-caused fires that don't meet the prescriptions, have been fought since their inception, but still burned more acreage. Prior to the implementation of this policy, all fires were fought, a policy grounded in the popular perception that fires destroyed forests, the wildlife therein, and the beauty of both. This was the policy in Glacier from 1910 until it was replaced with the prescribed fire policy. In many places throughout the West, this suppression policy (broadly used by the NPS and USFS) resulted in the accumulation of fuels that regularly occurring fires would have cleared out. Such fuels exacerbate fires that strike today, and while this occurs to varying degrees in Glacier, the important thing to know for now is that, since 1910, the majority of forests on Glacier's west side have seen fire. More importantly, since most of those fires were fought, nature's climate forces still have the upper hand in determining what forests burn in Glacier National Park.[39]

By pumping more greenhouse gasses into the atmosphere, humans are giving those forces ever more powerful hands. Not only are we creating the conditions for stronger and more numerous fires, but we're also changing the conditions for the forests that replace those that burn. Like Olympic, Glacier's summers are surprisingly dry, with only 13 percent of the annual precipitation falling in the three summer months at the Logan Pass level (12 inches historically in summer), and 23 percent at the Lake McDonald level (6.6 inches). If we continue business as usual, the Northwest Climate Toolbox predicts that annual precipitation will increase somewhat, but the increase will mostly come in winter and spring (partly as rain), while summer precipitation will actually decrease modestly. Consequently, the percentage of annual precipitation falling in summer will decrease to 10 percent at elevation (11 inches in summer by 2100), and 18 percent in the low elevations (6 inches). Along with hotter summers, there will also be less rainfall. Add in the earlier melt-off of the declining snowpack, and we will have created a seriously different environment for forests, especially at elevation, from what existed there historically. The same will be true for fire: a dramatically different—and more explosive—environment.[40]

What's more, an increasing share of the fires to come are likely to be big and extreme in behavior. As with Olympic, there's a good proxy for what such fires could be like in Glacier: the Big Burn of 1910, in northern Idaho and western Montana. That year began with a very dry winter and spring, and summer differed only in being hot. Thunderstorms in mid-July ignited some fires throughout the region, while sparks from coal-fired train engines and firefighting crews lighting backfires started others. By August, there were thousands of fires burning in the two states, including some in Glacier. On Saturday, August 20, a dry cold front moved into the region, packing hurricane-force winds that fanned the fires, combining several of them into four or five megafires that became the stuff of legend, stories, and history. Stoked by the wind, the megafires became firestorms miles wide, racing up the heavily forested valleys, jumping rivers, canyons, and mountain divides. Flames hundreds of feet tall created convection columns, pulling in more air to replace the scorched air gone skyward. In this way, the firestorms created their own weather, pulling in oxygen as the flames devoured trees by the millions, using all the available oxygen and making the wind blow harder. By the end of the day, a million acres had burned in the event's epicenter along the Idaho-Montana border, with another million the next day, and three million for the region as a whole. The ferocious wind subsided on the 22nd, and steady rain and snow began the next day, ending the Big Burn.[41]

Eyewitness accounts strain belief, with tales of cyclones of fire, forests being flattened by the screaming wind in advance of the flames, and entire trees being ripped out of the ground, lifted into the air, and ignited into flying blowtorches. Some saw whole hillsides explode at once, as pine pitch volatilized in the heat, becoming clouds of flammable gas that then spontaneously exploded, sending flames thousands of feet skyward and becoming a rolling wave of fire that destroyed everything in its path. Winds of eighty miles per hour created funnels, columns, and gyros of glowing, killing, incinerating destruction. Smoke and soot from the blazes darkened the skies in Denver and Saskatoon, Canada; enhanced sunsets in New York State; and even became a datable layer in the Greenland ice sheet. Firebrands rode the wind as much as ten miles ahead of the leading walls of flame, enabling the fire to leapfrog ahead of itself. Whatever the truth in these accounts may be, the human toll was significant. The infernos blasted right through seven small towns in their path, destroying them completely and taking out chunks of several more. Eighty-six people lost their lives, most of them firefighters. Some tried to escape on the last train out of town, but then took shelter in a tunnel after crossing a burning trestle. By far the most famous escape was that achieved by forester Ed Pulaski and his group of forty-three firefighters, whom Pulaski led into an abandoned mining shaft. When one of the men tried to leave the tunnel, Pulaski turned him back at gunpoint, almost certainly saving the man's life. Covering the tunnel entrance with a wet blanket, they rode out the firestorm, most of them passing out due to the extreme heat. Five of them breathed their last that day of horror, but the others all survived, most with first- and second-degree burns on exposed body parts. Pulaski was a hero to many, though he was never honored by the Forest Service. However, a firefighting tool he invented, a combination hoe and ax, fittingly bears his name still today, the Pulaski.[42]

The legacy of the Big Burn can be found throughout the West—and to a degree in Glacier—for it led directly to the "10 a.m. policy" that defined the USFS's approach to forest fires for the next half century. Under this policy, all fires discovered on the national forests were to be put out by ten o'clock the next morning. The policy gave the fledgling agency a raison d'être that Congress heartily endorsed—but it also slowly changed the West's forests, many of which had evolved with periodic fires. Without fire, not only did fuels accumulate on the forest floor, but the seedling trees that would otherwise have been thinned by fires also grew unimpeded, gradually closing in the previously open forests. When fires did eventually escape the 10 a.m. policy, they found a ready-made ladder into the canopy in the seedling-thickened understory. In this way, we unintentionally changed fire from nutrient recycler to destroyer on many western forests. As discussed above, the NPS pursued the same policy in Glacier, though the adverse effects are not as pronounced there. Another part of the Big Burn's legacy is its status among the country's big fires. Even with the terribly destructive fires of the twenty-first century, the Big Burn is still considered by many to be the worst in American history. That it happened before the effects of climate change were beginning to be felt should give us pause. If it happened then, in a region that is 11 degrees cooler than what our grandchildren will likely face at the end of this century, what will such fires be like then?

Glacier's recent fire history demonstrates that ecological and ice resources aren't the only things at risk of damage or loss due to wildfire induced and enhanced by climate change. In 2017, the Sprague Fire burned down the dormitory at the Sperry Chalet, a backcountry lodge built in 1914 by the Great Northern Railway. The following year, the Howe Ridge Fire destroyed several historic buildings around Lake McDonald, including the main lodge at Kelly's Camp (a historic tourist cabin resort) and some outbuildings, and the boathouse at a summer home for Montana's well-known US senator Burton Wheeler. Fires can also destroy anything organic at archeological sites, thereby compromising our ability to learn about past humans and their lifeways. As with most national parks, only a small portion of the archeological sites in Glacier have been surveyed, so such damage causes an irretrievable loss of knowledge. Wildfire is just one of the ways that climate change can damage or destroy cultural resources; enhancing the potency of extreme weather events can cause widespread damage through more severe winds, floods, and storm surges. One need only look at the hurricanes of the twenty-first century to find examples of such destruction. Through wildfire and more powerful storms, then, climate change threatens irreplaceable parts of our cultural heritage.[43]

At Lake Frances, once Woody and I had our fill of goofy pantomimes, we dried off and retraced our steps to Thunderbird Pond, where we cooked and ate our dinner. Enjoying the tranquility there, we were quiet for a moment, listening to the sounds around us. The evening was still, with so little air movement that the pond perfectly mirrored the vaulting mountains above it. An occasional bird chirped among the willows lining the shore, but otherwise there was no auditory indication of animal life. Filling our ears, though, was the distant sound of Thunderbird Falls, the background music to the regal scenery all around us. The same subdued music had been playing at Lake Frances, where the meltwater stream from Dixon Glacier soothed the soundscape (everything one can hear from a single location). It had been playing the whole time we'd been at Hole in the Wall and on the climb out of the Bowman Creek Valley, both of which also had meltwater streams cascading down distant mountainsides. The background music isn't heard everywhere in Glacier, as I had found at Goat Haunt, but it's common enough that it's an integral part of the soundscape of Glacier's high mountains (see plate 18). And while it won't disappear as climate change tightens its grip on the park, it will get harder to find, especially in summer, after the winter's snow has melted away. This unique soundscape of Glacier's mountainous core is yet another thread that climate change is unraveling.

Woody and I finished our dinner and cleaned up, then climbed the hill back to our campsite. The next day, as we began the trip back to jobs and doctors, we had a grizzly bear sighting a few minutes into the hike. A sow and two cubs were foraging in a meadow kept open by avalanches, two or three hundred yards from the trail. They seemed unaware of us, so we watched them for a while. It was a capstone of sorts for the trip, the pinnacle of wildness doing what comes naturally. The bear sighting was also a reminder that we shouldn't lose hope, for as an adaptable generalist, grizzlies will most likely persist through the changes to come. They have already survived the loss of whitebark pine nuts, the trees having succumbed to white pine blister rust decades ago in Glacier. The pine's nuts are high in fat and protein, and are borne in fall, so they were the perfect food for the bear to prepare for hibernation. Today, huckleberries are a mainstay for them here, the sweet fruit providing plenty of calories in most years and constituting 15 percent of their annual diet. Lest we be too sanguine about the changes to come, though, we don't know how huckleberries will fare as climate change really sets in, so any efforts we make to abate global warming's worst effects just might help huckleberries persist, along with the bear. Indeed, if the past is any indication of the future, our own actions can help save the bear, just as our predecessors' actions gave us Glacier National Park. The bear is not the only reason to have hope—but we must act.[44]

Like the bear (hopefully), Glacier's remarkable landscape will always inspire, even if the worst climate change effects are realized. Those shining mountains will outlast us and our climate change–induced damage, which gives me a small measure of comfort when I despair over our seeming inability to address this threat. The same is true of the Grand Canyon and Olympic: these are sacred landscapes, which will always inspire, to varying degrees. But the more we tarnish their beauty and alter or destroy their primeval forests, the less they can impress and amaze. For this reason, we need to act *now* to stave off the worst of runaway climate change. Who knows? We might even save one or two of Glacier's glaciers—along with the sound of falling water.

The Hole in the Wall trip was the last time I saw Woody, for his health soon began the decline so typical of cancer victims; he passed away the following July. I saw Janet many more times, beginning with a trip to scatter Woody's ashes a year after he died, on his favorite mountaintop in California, and ending with the 2014 Goat Haunt trip.

The Hole in the Wall trip complete, let's rejoin Janet and my Yosemite friends Charlie and Margaret at Goat Haunt. Janet and Margaret had left Charlie and me for a hike to Lake Janet; they returned midafternoon happy but with some mosquito bites, despite wearing long sleeves and having used repellant on exposed skin. With them in camp to keep me company, Charlie took off for some exploring in the immediate area. When he returned, we spent an enjoyable evening together, under partly cloudy skies. After the last visit by the *International*, we had the pavilion to ourselves, and once again found the deep quiet of the Glacier lowlands, the perfect complement to the high-country soundscape Woody and I had enjoyed not so long before.

The next day, Janet and I left on the first *International* trip out, while the Yosemite folks stayed behind for more exploring. Saying goodbye on the boat dock, Margaret had tears in her eyes, knowing we might not see each other again. I felt numb, wanting to believe otherwise and unable to contemplate a very different future for myself than the life I had known. It was denial—but then, I don't think any of us would want to ponder a future with ALS in it. Whatever the case, we had to be going, so we boarded the boat, waved goodbye to our friends, and left Goat Haunt, taking our memories with us. After lunch at the historic Prince of Wales Hotel (situated on a prominent hilltop above the townsite and lake), Janet and I drove back to Gardiner. After helping me unpack completely (she knew my weakened hands and shaky balance would make that difficult), she left, also with wet eyes. My numbness was unchanged; just nine months earlier, I had visited her on an extended, solo road trip through Montana, Wyoming, and Colorado, in excellent physical condition despite my recent diagnosis. What a difference those months had made—not even a year!—destroying much of my walking ability, speech legibility, and hand strength. I had gone from living independently to needing assistance with most everyday tasks, a trend that would continue throughout the rest of the summer. It's no wonder I was numb to such situations and thoughts, for it was my way of staying sane.

As July turned into August, my arm and upper-body strength had eroded so much that I struggled to lift a plate of food into the microwave over the stove. Weakening fingers were a constant nuisance as well. Dressing myself was a daily challenge, frustration, and time sink, often taking an hour when combined with a shower. Dropping things was an almost daily event; picking them up, a Herculean challenge that I increasingly asked Pete or friends to do. Tasks that required both arm lifting and finger strength, such as trimming my beard with scissors (my technique for over twenty years), became more difficult, and eventually impossible (so Sean added barber to his list of skills). More seriously, I could not always get up from a fall. Twice, I fell at home with no one else around and could not get myself up. The first time, I had no choice but to wait till until Pete came home an hour later. Imagine the thoughts and despair that go through your mind while you wait, totally helpless on the floor, for someone to arrive. Has it really come to this? Why do I have this awful disease? What did I do, or not do, that saddled me with the worst-possible neurological disorder? And so on. The second time, I used the emergency call service that friends had suggested I get after the first solo fall; by pressing the button on the bracelet I wore, emergency response was triggered and help arrived quickly, but not before I again resigned myself to physical failure. The body that had once carried a fifty-pound backpack for a hundred miles of rugged mountain terrain now could not get up from the floor, or even bend down to retrieve a dropped coin. The disease was progressing alarmingly fast.

Of all the disease-related things that happened to me that summer, by far the most indelible was an altercation with Pete that expedited both of our departures from Yellowstone. Pete was a fellow that my parents had met while they were helping me move to Gardiner in May 2014. Gardiner, like many national park gateway towns, has too little affordable housing for the people who live and work there, especially in the summertime. Pete needed housing for the summer and seemed like a reasonable guy, and I increasingly needed assistance in the house and had extra bedrooms, so we all could see the benefits of his helping me around the house in exchange for reduced rent. So we made a handshake deal, and he moved in when my folks left for Missouri a few days later. And, for the first half of the summer, this arrangement worked out well—though I found myself wondering about his background. He was fifty-one, yet he worked an entry-level job at a nearby hotel, had almost no savings, and didn't even have his own vehicle. He told me that he was just emerging from a low point in his life, which we have all experienced, but certain things he said kept me wondering if there was something else at play. For example, he told me that his previous job had been delivering phone books in rural Wyoming, another entry-level job. There didn't seem to be a career path in his past, which was not a problem for our situation, but I continued to wonder why.

Midway through the summer, he met a girl. I remember thinking, when he first told me about her, that this could spell trouble, for I knew that the euphoria of love could distort his thinking and perspective on life and its challenges. Their very relationship was evidence of this potential problem, for she was just twenty years old and was only in the country till mid-September, when she had to return to her home and college in eastern Europe. Naturally, the two of them didn't see a problem with their age disparity, but her impending departure was always in the background for them. The visa under which she could work in America could not be extended, and Pete didn't have the financial discipline or income to afford the airfare to accompany her home. So Pete began focusing on the causes he believed responsible for that problem: his low-wage job and his landlord. Initially, the hotel at which he worked was his primary focus, and his relationship with his boss quickly deteriorated. By the end of August, she had fired him for insubordination, which both exacerbated his financial situation and freed him to focus his attention on me. As it happened, I was away at the time, which gave him time to brood about his situation. I know this in hindsight, but I was about to unwittingly discover the depth of his fury—and the probable cause of his poor job history.

My time away concluded with a visit from my friends Ann and Mark, who went biking with me near Gardiner. We had lunch together and then went to my house, where they began to gather their things in preparation for the four-hour drive back home. My first clue that something was up with Pete was seeing him leaving the kitchen with a beer, which he told me was not one of mine. Pete had told me previously that he and alcohol didn't go well together, so it appeared he had fallen off the wagon. He returned a couple of minutes later, and I tried to stimulate a conversation by asking him how things were going. His reply was both odd and revealing: he said that he loved Maria, his girlfriend. I knew that already and was tired of his pity party regarding her and their ill-conceived romance, so I replied by asking him if he was planning to go see her in Europe, which is what most people I knew would do in his situation. He angrily asked me where he would get the money for that, his face contorted in an ugly sneer. He stormed out of the room but soon returned, asking me the same question, again with the sneer—and again with both Ann and Mark present, witness to it all. He left the room again, only to return yet again, one or two minutes later. I don't remember what he said then, but the three of us could see a pattern developing, which then carried on for more than an hour. Time after time, he would climb the stairs to the living room, hurl some invective at me, and then leave the room. Ann and Mark, who really needed to get going, reassured me that they were not about to leave me alone with him. Each of them followed him out of the room to tell him that he'd made his point, and that he could leave town, if that's what he really wanted. But instead, he would come back for more shouting.

As the hour wore on, it became clear that Pete had lost control of himself and could not contain his rage. Instead, he was becoming more and more agitated, using four-letter words, standing closer to me, and adopting an increasingly threatening posture. I didn't give him eye contact, which probably kept the situation from spiraling completely out of control. But I knew that he owned a pistol, and began to fear that, in his state of mind, he might bring it into play. Wanting to prevent a real accident, I began urging my two friends to call the sheriff. I told them why—that he had a gun—but with my slurred speech, they couldn't understand the pivotal word (I couldn't say the hard "g") and in the heat of the moment, I couldn't think of synonyms like pistol or revolver. Finally, I used my iPhone to spell it, at which point Ann's eyes widened noticeably. I also had them call Sean and Missy, who were away that weekend. They urged Ann to do the same thing, so she did. About that time, Pete actually left the house, telling us (threatening us?) that he would be back in an hour. The deputy sheriff arrived while he was out, and we told him what had happened. He told us that, a week ago, he had escorted Pete from the hotel at which he'd worked when he became belligerent toward his former boss. At that time, the deputy had wondered how Pete would be with his landlord and roommate; now he knew. When he asked me what I wanted to with Pete, I said to evict him, knowing I would never be able to sleep comfortably with him in the house. After some more discussion, the deputy said he would see to it.

With that, the drama was basically over, though Pete did ask me why I was not giving him the month's notice that he and I had agreed to. The deputy replied for me, telling him that it was his behavior, to which

I nodded my agreement. Pete also called his mother, who had visited him just days before. She asked to speak to me, and rather than trying to persuade me to let him stay, she apologized to me and told me he'd had anger management problems since he was three years old. I didn't ask her if she had pursued any remedies for his problem while he was a child, and many times since, I have wondered if more accessible mental health care would have enabled him to effectively manage his problem. Ann and Mark left shortly thereafter, and Pete was gone within a few hours. The locksmith down the street changed the locks that evening, and neighbors and friends kept me company until Sean got home. He spent the night with me, a night that gave little rest, given the intensity of the day. Again, I had felt completely helpless, a feeling I would come to know intimately. My father flew in the next day, but he couldn't stay with me into October as I had planned, so I found other friends who could stay with me through its first week, which allowed me to attend a conference in Yellowstone that I had always enjoyed. The day after it ended, I flew home to St. Louis, two weeks earlier than I had desired. Those two weeks saw some of the best fall weather in recent memory in Montana and Yellowstone, which I would have relished.

One of the friends who helped me out that last week in Gardiner, Steve Swanke, took me to the Bozeman airport that October morning. Two other friends, Doug Hilborn and Sean, came to the house to see me off before we left. Doug is a funny, lighthearted guy, and he told me he would not tolerate any tears. That was a prescient, convenient remark, for it gave my numbness cover. We said our goodbyes, and I got in the car to leave. I tried to get one last look at my would-be retirement home, but I didn't get the view, for I was on the wrong side of the car. Two hours later, we were in the airport terminal. When the time came to board, I saw that Steve, a retired law enforcement ranger accustomed to containing his emotions, had wet eyes. I was, once again, numb to the reality of what this departure meant. This time, there would be no coming back, no retirement, no joyous return, but I couldn't admit that to myself. Instead, I remember thinking (or hoping) that I might somehow be able to return for another summer. Wishful thinking it was, for even at the time I was having great difficulty stepping out of the bathtub, but it was my way of coping. Hope springs eternal, which was certainly true of myself that day. I think it still does, despite the experience I have gained with the disease since that bittersweet summer.

Looking back on that summer, I have two dissimilar impressions. The disease progression brought an increasing number of struggles, a growing list of difficulties that detracted from, and almost came to define, my final season in a beloved place. I had envisioned a relaxing and fulfilling time, but what I got was a season marked by frustration and adversity, at least partly. From the daily challenges that were more or less mundane to the exceptional challenge of dealing with a man unable to control himself, the summer was marked with emotional difficulties (while few ALS sufferers have an incident as intense as mine, many do deal with other difficult situations, including theft and abuse by home health-care workers and other caregivers). The other impression, though, is one of peace and tranquility, warmth and community, amazement and happiness. The Glacier trip embodied these impressions as much as anything I did that summer, for it brought three good friends to Montana to take me to a park that I loved almost as much as Yellowstone. There we found the deep quiet that can only be found in nature, we enjoyed the conviviality of friendship, we relived fond memories, and we were amazed by the landscape around us. We made new memories that sustain us, memories that I draw upon every day, especially when I write passages like this chapter. I have not been back to Yellowstone or Glacier since Steve took me to Bozeman—or any national park, for that matter—but I continue to feel strength from these experiences.

The memories I have from Glacier (from all seventeen trips I took there, not just the two discussed here) are perhaps, as a group, the most vivid, even sensual, of all the memories of places in which I have spent significant time. They are an artist's palette of bright color, not only from the flowers that grow in such abundance there, but also from the blue lakes and sky, the mountains tinged in hints of green and red, and the deep-green conifers. My memories—and dreams too—carry these dimensions with them, from mountains soaring into the sky to waterfalls dropping back to earth, from the tiniest alpine flower to the shoulder-high bear grass that colors your shirt with pollen when you brush by, and from the powerful grizzly that reminds you that you're the visitor here to the diminutive pika that greets you when you sojourn to its alpine home. Clearly, the dreams and memories are not only visual, for I also frequently hear the sound of falling water, feel crisp morning air, and smell the scents of flowers and trees. The place is a sensory overload, even in my dreams and memories. Glacier is life, and it gives me life still.

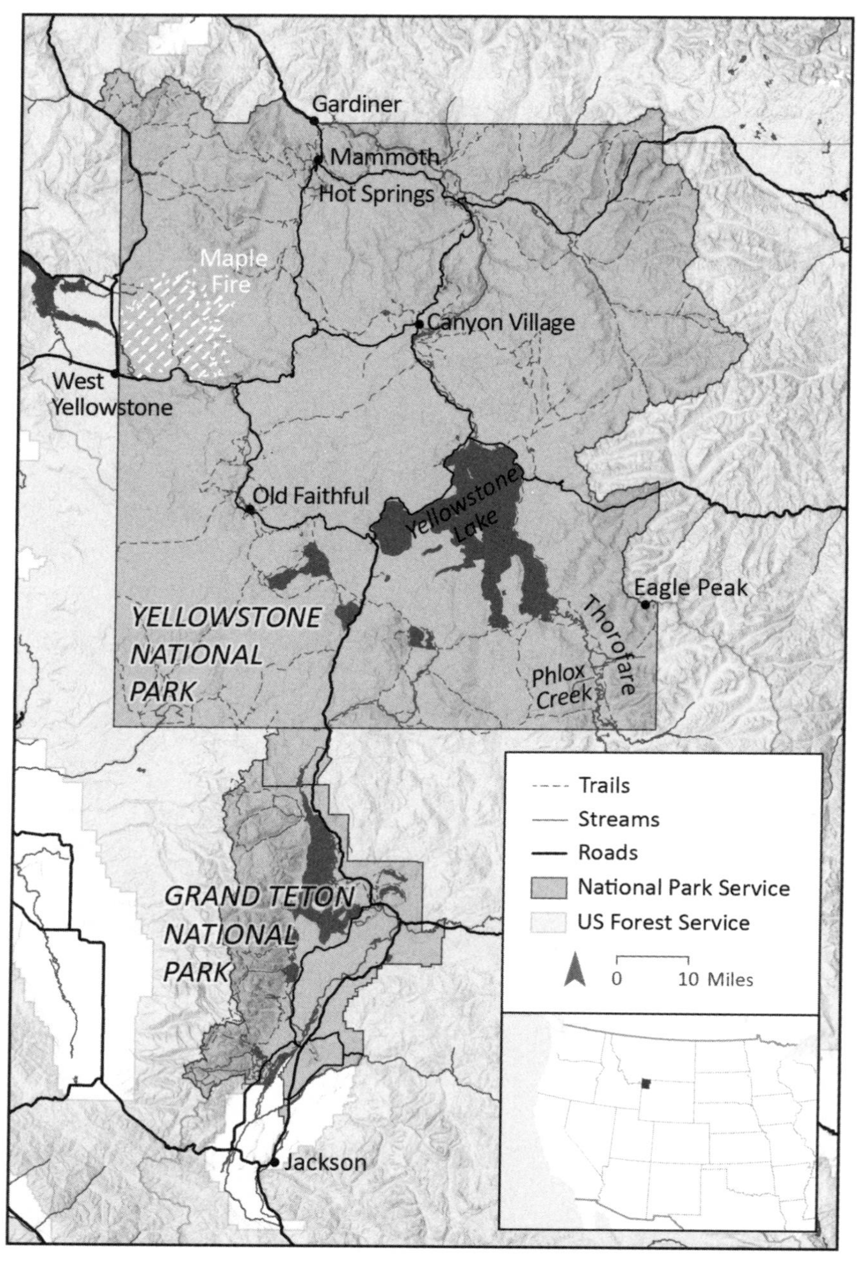

MAP 4 Map of Yellowstone National Park.
Sources: NPS, USGS, Natural Earth

Chapter Four

YELLOWSTONE NATIONAL PARK

Yellowstone National Park is a land of superlatives: The first national park in the nation and, by some definitions, the world, and progenitor to thousands of national parks in over a hundred countries.[1] The world's largest collection of geysers, hot springs, and other thermal features, with more geysers than the rest of the world combined. The largest lake in North America at its elevation, Yellowstone Lake, at 7,733 feet above sea level. The home of America's only herd of bison that has continually ranged freely in the wild—and the purest stock of bison genes in the country. The heart of the Greater Yellowstone Ecosystem, one of the largest, mostly intact ecosystems in Earth's temperate zones (about 28,000 square miles, of which Yellowstone is one-ninth). One of only three parks (and surrounding ecosystems) south of Alaska that harbor three of the continent's top predators—grizzly bears, cougars, and gray wolves (the other parks being Grand Teton and Glacier). The epicenter of one of the largest calderas (collapsed volcanoes) and active volcanic hot spots in the world. The location of the coldest recorded temperature in the coterminous states from 1933 to 1954, -66 degrees (eclipsed by Rogers Pass, Montana, at -70 degrees, a record that still stands). The location of one of the largest wildfire events in contemporary times, burning almost 800,000 acres in 1988 (36 percent of the park). Finally, a 2.2-million-acre fastness in the northern Rockies, with 1,386 native plant species including three endemic plants, two of which can only be found in the park's thermal areas.[2] For these reasons—and many more—Yellowstone was dubbed Wonderland in its early days, a moniker that only seems more appropriate the richer our understanding of the park becomes.

When I first began working in Yellowstone in the summer of 1986, I knew little of these distinctions. I had been to the park twice, once as a two-year-old and again as an impressionable teen. The initial visit gifted me and my twin, Jim, with our earliest childhood memories—of our dad wearing sunglasses for the first time (in our limited experience). Even as toddlers, we noticed how important they seemed to be, which may have been the reason we both erupted in tears when a sudden gust of wind blew them off his face and into the Grand Canyon of the Yellowstone, never to be retrieved. Twelve years later, we returned, this time with two younger brothers. Taking some ranger-led nature walks, we encountered the eclectic people who run the parks. One ranger let a mosquito bite her, claiming immunity to their toxins and using the pesky incident to teach us why they bite and their role in nature. Another ranger encouraged us to get on our knees on a floating boardwalk across an estuary to see all the life below us. The floats supporting the boardwalk were not intended for a dozen or two stationary people, so we got our knees wet, but made another memory featuring Yellowstone and its distinctive stewards. Wanting to know how I could join their ranks for my upcoming college-year summers, I received a pamphlet with a list of addresses not only for the NPS but also for the companies that ran the hotels in the parks. I applied to several parks and concessioners, but the only position I was offered was an entry-level summer job in the Canyon Lodge gift shop in Yellowstone. Still, it was a foot in the door—and another step in my inauguration into Yellowstone's wonders.

Like most new employees, I spent that first summer visiting Yellowstone's well-known attractions: the geysers and hot springs, canyons and waterfalls, lakes and mountains. I also began exploring its immense backcountry, that two-million-acre wilderness away from the roads, where human development is absent and nature is on full display. I climbed mountains that had never been mined or logged or carved into ski resorts, I paddled lakes that had never been dammed to raise their level and store water for downstream irrigation, and I explored backcountry thermal basins the likes of which Iceland and New Zealand have turned into hydrothermal energy plants. I marveled at a natural abundance that, for the most part, is the stuff of history books and frontier explorer journals.

Finally, I witnessed wild animals going about their lives, heedless of our existence, and natural forces like fire march across the landscape, again heedless of our existence—or our attempts to corral them. Moreover, it wasn't just Yellowstone that was this way, but the millions of wilderness acres surrounding it, all protected from our sometimes tawdry development by people who recognized that there was something special there. Mesmerized by that enchanting landscape, I returned for a second summer, then a third, and eventually two decades of my life. Like many who have spent time in Yellowstone, my life was forever changed by that first summer and the many that followed.

When I arrived in St. Louis in October of 2014, then, I decided to write a book about the deeper meanings and values of Yellowstone's amazing landscape. Reflecting on my twenty-two years there and a wealth of encounters with Wonderland, *A Week in Yellowstone's Thorofare* focuses on three threads uniting our collective experience with Yellowstone: beauty, community, and wildness. Beauty is perhaps the most obvious thread; it's something we have long desired and appreciated. Indeed, beauty is so fundamental to humans that it's one of thirty-six values common to all of us.[3] Whether we find it in art, gardens, or the natural world, our call to seek, create, and appreciate beauty is as old as humanity itself. Community is possibly the most surprising thread, especially for someone who has found Yellowstone's meanings most evident in the backcountry, away from people. But here again, our need for each other is one of the things that make us human. We are one of the most social animals on the planet, and setting Yellowstone or any natural area aside is almost always done for society's benefit. Moreover, we usually experience Yellowstone with others, whether in the car or miles away from the nearest road. Even solitary trips into wilderness are often done to reflect on our role in the circle of people we go through life with. Last, wildness is perhaps the most subtle thread, at least at times. But the longer I was in Yellowstone, the more I saw it everywhere. It's in the free-flowing rivers, the untarnished mountains, and the naturally erupting geysers. It's in the bugling bull elk, the roaring grizzly bear, and the serenading trumpeter swan. And it's in the superlatives: the grandeur of a sunset over Yellowstone Lake, the drama of a wolf pack taking down a bison in Hayden Valley. Yellowstone is beauty, community, and wildness—and a whole lot more, to me and millions of other people worldwide.[4]

As with chapters of this book, *A Week in Yellowstone's Thorofare* embedded these concepts within a wilderness trip. In that case, it was my last wilderness trip, in August 2014, when three good friends spent a week paddling me through the Southeast Arm of Yellowstone Lake. In doing so, we skirted the edge of the Thorofare, a two-million-acre wilderness area composed of Yellowstone's southeast corner, the Washakie Wilderness in the Shoshone National Forest, and the Teton Wilderness in the Bridger-Teton National Forest (see plate 19). The curious name Thorofare comes to us from the fur-trapping era of the early to mid-1800s, when a route within that wilderness became the easiest way for trappers to get from the Jackson Hole area onto the Yellowstone Plateau. It became known as the trappers' thoroughfare, and the abbreviated version of the name stuck. During my early years in Yellowstone, I coveted a trip into the Thorofare, but didn't know how to do a weeklong hike (the minimum time most people need) without breaking my back carrying an eighty- or ninety-pound pack. Once I learned how to minimize the weight, it was a few more years before I figured out how to make the best hiking conditions align with a week off in the busy summer season, but I eventually figured it out and became a repeat visitor there.

I chronicle a few of those earlier trips in both *A Week in Yellowstone's Thorofare* and my next book, *Essential Yellowstone*.[5] For this chapter, then, let's journey into the Thorofare wilderness again on a trip that did not make it into either book, but one that featured a climb to the summit of Yellowstone's highest peak—and that showed us signs that the park's climate was already changing. The passage of more than a decade has only amplified the altered conditions that threaten some of the very superlatives and meanings for which the park was established to protect in the first place.

Standing on Terrace Point on the shore of Yellowstone Lake and looking south into the Thorofare that late August day in 2006, it was hard not to be impressed and excited. Sprawling in front of Josh Becker (whom we met in the introduction) and me was the Yellowstone River Delta, the pancake-flat area where the river slows before entering Yellowstone Lake. The Southeast Arm of the lake, as blue as the Rocky Mountain sky, also sprawled, but behind and to our right. The delta was covered in willows that practically glowed green in the bright morning sunshine. Here and there were sprinklings of firs and pines trying to make a go of it in the delta's frequently saturated soils. In front of us were shallow lenses of water, the remnants of creeks that also flowed into the lake, or puddles the Yellowstone River had not yet filled in with the sediments it drops upon entering the lake. Various ducks swam about the lake's shallows, the water just a few inches deep there, perfect for foraging. The river itself was lost in the willows that extended two or three miles to our right, the greenery ending where the Two Ocean Plateau abruptly began rising two thousand feet above, in shades of indigo and umber. Following that alpine plateau to the left, we could just barely see where the delta's willows and wetlands give way to forests and grass, the vegetation of drier soils. Beyond that and further up-valley, the blue northern ridge of the Trident formed the distant horizon, its alpine spear ascending gently to the left. Further left and returning to us was Colter Peak, rising in bumps and lumps from the valley, eventually topping out two or three thousand feet above the flatness. It was a lively, colorful view of the best of Yellowstone and the northern Rockies, one that epitomized the beauty, wildness, and community I have been fortunate to find throughout the Yellowstone region.

It was also a view into Yellowstone's past, for it included much of the fur trappers' thoroughfare. Looking again to the farthest end of the valley some ten or twelve miles away, we could see a notch where the plateaus on either side of the valley seemed to pinch close, perhaps forming a narrows through which the river passes. The notch was an artifact of our vantage point, for it does not really exist, but is an optical illusion marking the place where the valley turns slightly. However, it did guide our eyes to the trappers' thoroughfare, which came through the imaginary notch, proceeded to the delta, then continued around the lake in either direction. As if it were his watchtower, the peak named for the original "mountain man," John Colter, looked over all of this, the northern half of the fur trappers' thoroughfare. Colter left the Lewis and Clark Expedition on its way home in 1807, to join some trappers exploring the new Louisiana Purchase to capitalize on the West's riches. He was probably the first Euro-American to see Yellowstone's wonders, decades before those were known to contemporary Americans—and he did it in winter, in an area that routinely saw temperatures dipping to -40 degrees! While he did see Yellowstone Lake, he probably did not come to the lake's southern arms, or, therefore, the trappers' thoroughfare.[6] Nonetheless, it seems fitting that a mountain in the Thorofare bears his name, for the region still looks and acts much the same as it did for him. A few trails and patrol cabins now facilitate travel on foot or horseback, but it's still a place rich in nature's bounty, two centuries later.

That wild land with the incongruous name would be ours to explore for the next week. Terrace Point was our changeover spot, where we stashed my canoe and shouldered our backpacks, just like Woody and I would do in Glacier the next spring. Josh and I had paddled down the east shore of Yellowstone Lake the previous morning, some thirteen miles in my canoe. We'd left my home in Gardiner well before dawn to get an early start on paddling while the lake was calm. Afternoon breezes can turn the lake into whitecaps, especially its east shore, at the receiving end of fifteen miles of open water aligned with prevailing wind. On the lake by seven that foggy morning, we had nothing but stillness, so the paddling was easy and efficient. After a couple of hours of paddling, we stopped at Park Point for a snack, to stretch our legs, and to climb the grassy hill for a peek into our immediate future. The sky was overcast with high but thinning clouds, and with no wind, there was nothing more than ripples on the aquatic trail we'd been following. The Southeast Arm was beckoning, so with our energy stoked, we resumed paddling. Just in time for lunch, we pulled into our campsite on the arm, centered on a small clearing. The site looked across to the Promontory (the low ridge separating the South and Southeast Arms), and beyond it, the Red Mountains. We set up camp and ate, content with the day, and the next day had another few hours of calm paddling before arriving at Terrace Point by midmorning. It had been an excellent trip thus far.

Being out on a lake as the day begins is a true delight, involving all the senses. The eyes see tendrils of fog rising from the treetops, while the ears hear nothing but the sound of water dripping off the paddles as they're brought forward for the next stroke. The mouth and nose detect dampness and wet pine needles, while the skin feels the chill and humidity of the rain that fell last night. Staying close to shore, we enjoyed the closer view of damp forest, the organic silence, the rippled reflections, the feel of upper-body muscles working. We spoke with hushed voices, not wanting to divert attention from the all-senses show through which we canoed. Occasionally, a shaft of brightness way to slumber (at least it broke through the shoreline forest, hinting that warmth might be coming. The sun strengthening, the fog gradually lifted, revealing distance, lake, horizon, sky, clouds. The sensory experience became more visual with this second act, though the others still registered the impressions of warmth and exertion, water and gentle breeze. Once we landed at our campsite, the afternoon brought act 3, as a thunderstorm loomed, hit, and moved on. We watched the southern sky fill with darkness and clouds of potency, moving toward us, subsuming the sun, and flashing with lightning. Gusty winds picked up, bringing scents of rain and electricity, falling temperatures. Thunder and rain arrived, first a few heavy drops that splattered into aquatic silver dollars, then a full-volume staccato on the tents. The rhythm softened, then roared again, quickly accompanied by waves pounding the cobbled beach. Quickly, the storm moved on, quieting the downpour, but the pitter-patter of gentle showers lingered for more of the afternoon, as did the wave action. The evening brought clearing skies, coolness, and dampness, and a riot of color in the western sky as day gave way to night. The sun itself sunk into a low point in the Promontory, and the purples and indigos of dusk and night took over the land. The all-senses experience gave us the promise of more to come.

The next day was equally sensual, beginning with a reprise of the fog. We had only an hour of paddling before it exited the scene, by which time we arrived at Terrace Point. Stashing the canoe, we dried the tents in bright sunshine, shouldered the backpacks, and headed south on the Thorofare Trail. We hiked beyond the delta and into somewhat drier forests interspersed with a liberal assortment of meadows large and small, moist and dry. The scents and feels of humidity and dampness gave way to the dry spice of pines and grasses, warmth and sun. At Beaverdam Creek, we numbed our feet splashing through the edge of, appropriately enough, a beaver pond that had covered the trail for a few dozen yards (rodents that can read a map, who knew!). Passing the unoccupied Cabin Creek Patrol Cabin (humans that adhere to place-names too, who knew!), we took lunch on its solarium porch, enjoying the warm-up that our cold-blooded neighbors need to survive on spaceship Earth. Hiking another three miles, we set up camp at the base of Turret Peak, a rocky pedestal guarding a treasure—the wild Thorofare. The mountain was a blend of brown and purple, dark against a white thunderhead glinting brilliant in the blue of the late summer sky. Complementary colors defined the foreground: grasses cured golden brown; willows exuberantly green; sedges a marriage of green and yellow; pines the forest green expected of them; and, in the midground, a swath of fire-killed trees, balancing the vibrant colors with their neutral gray. The evening brought sounds that told us definitely we were not in the city anymore: the hoot of an owl, the bugle of a bull elk, the howl of wolves. Few things speak of nature's mystery more than a great horned owl hooting, its call seemingly floating to us through the darkening twilight. Likewise, almost nothing evokes the feel of autumn in the Thorofare more than a bugling elk, the oddly high-pitched song conjuring frosty mornings, Indian summer cobalt skies, golden willows and red fireweed leaves. Finally, nothing else conveys—virtually the world over—the sense of wildness as the call of the wild: wolves howling into the night. We had arrived in the Thorofare at the cusp of autumn, the big wild.

The next morning, we found the tracks of the canine a cappella singers that serenaded us overnight, along with grizzly bear tracks—both of them fresh. It wouldn't be a Thorofare hike without grizzly tracks, for the area is excellent habitat for them, and I have seen their tracks every time I have hiked through the area. On several occasions, I saw the track makers themselves, but never faced a charging mother bear or a silvertip boar intent on procuring his next meal. I have related the stories of those encounters in my previous two books, so I will sum them up by saying they are always gripping, even if there is little chance of injury.

The possibility of an encounter is always in the back of one's mind in a place like this, which gives the Thorofare an added measure of wildness. Wolves and mountain lions, the other two top predators in the Yellowstone region, have both been known to kill people (wolves primarily in Europe), but they don't have the power that grizzlies hold in our imagination. This is probably due to the fatalities that grizzlies have caused, especially those few in which a grizzly stalked and killed a person to eat them. Also contributing to their image are some of the stories survivors of bear attacks have told, along with the emotions the stories kindle in their listeners. Regardless of the reasons (there are many others), the presence of the bear heightens the senses, demanding that hikers be aware of the landscape through which they travel. This need only enriches the sensory experience of the Thorofare.

Like many areas of Yellowstone, large parts of the Thorofare burned in the 1988 fires. The Mink Creek Fire, the main one to move through the area we were in, began with a lightning strike on July 11, in the neighboring Teton Wilderness. Park managers, believing the typical wetting summertime rains would reappear as the National Weather Service was predicting, allowed the fire to burn into the park. They had made similar decisions on a number of lightning-started fires already that summer, and the majority of those fires had gone out on their own, as they had done in most of the years since the park instituted the prescribed fire policy in 1972. This time, would be different, as the summer would go on to become the driest and warmest in Yellowstone's history (records that still stand). In July and August, a series of cold fronts passed through, most packing wind and lightning, but little or no rain. The summer had two climaxes, the first occurring when the most potent cold front blew the flames across 240,000 acres in one day, doubling the acreage burned thus far. The day became known as Black Saturday, an eerie twist of history in that the Big Burn also had a Black Saturday in 1910—on the same date, August 20, when 1,000,000 acres burned. This uncomfortable historical reprise did not extend to Yellowstone's second climax in 1988, when the North Fork Fire blew through the Old Faithful area, luckily sparing the historic Old Faithful Inn. The turning point in that summer of fire came on September 11, when a modest amount of rain and snow fell in the park. It wasn't much, but it was the first significant precipitation park-wide in three months. The heavier snows of October and November put the fires out for good.[7]

Over the following winter, park staff members produced all sorts of assessments and inquiries, as well as an approach to help the public (and themselves) understand what happened and why, and how nature would respond.[8] The story that emerged from those efforts was based on recently published research done in Yellowstone, and the serotinous cones of lodgepole pine figured prominently in the story, along with the particulars of Yellowstone's climate and fire history. Large swaths of Yellowstone, especially the park's high-central and southern volcanic plateaus, are covered in lodgepole pine. These forests consist of a patchwork of even-aged stands of trees, with most stands dated to the mid-1800s or early 1700s. This evidence indicates that the lodgepole-dominated forests of Yellowstone burn every one hundred to four hundred years, that those high-intensity, stand-replacing fires open the serotinous cones and start a new forest, and that it takes another one to four centuries for sufficient fuel to accumulate for the next round of fires. Importantly, though, according to a key paper on this subject (published in 1982), such fires would not be possible until "warm, dry, windy weather" occurred.[9] That weather arrived in spades the summer of 1988, and the fuels to support crown fires were present throughout the park. Consequently, fires burned 36 percent of Yellowstone (over 1,000,000 acres, when the acreage burned in the national forests surrounding Yellowstone is added to the park's 793,880 burned acres), and seedling lodgepoles took root in the newly exposed soils.

Indeed, that story helped us all make sense of what seemed to many Americans a tragedy. The fires were nature's way of renewing itself, doing what it had been doing for millennia. This was outside of the popular ecology we had learned from films like *Bambi* and characters like Smokey Bear about the damaging role of fire in naturally functioning forests. But it was true nonetheless, and once the seedlings grew tall enough to see from a moving car, most visitors accepted it. I myself told the story in my park ranger campfire talks for three summers, and I took visitors to a site just off the West Entrance Road, where seedling lodgepoles were so thick a person couldn't walk through them. We called such a forest a "dog-hair stand of trees" because they were as thick as the fur on a dog's back. It was a site that probably burned on the short end of the fire intervals, about once a century, and illustrated the extreme of lodgepole pine serotinous fecundity. In places where stand-replacing fires are regular events, up to 80 percent of lodgepoles bear serotinous cones, a textbook example of adaptation to local site conditions. (Not all lodgepoles bear serotinous cones; throughout North America, more of them actually bear the typical pine cone that opens normally, without fire.) It's a marvelous and powerful natural cycle that we were fortunate to witness some thirty years ago—more powerful than even the federal government, which staged its largest fire-fighting effort to date that year, but was largely defeated in the backcountry (firefighters had much better success in the front country, losing only a few structures).[10]

Researchers from all over the country recognized a unique opportunity to study the various aspects of fire ecology in a wildland setting, producing dozens of published research papers on all aspects of the fires and nature's cycles and recovery. By far, this research thrust supported the fire story; some of the more interesting findings follow.

First, many scientists looked at the recovery and succession of plants in the burned areas. Revegetation occurred quickly, with the understory vegetation equal in amount of cover to that before the fires by 1996. Few invasive weeds were present, and the soil's nitrogen and other critical plant nutrients remained in place (in contrast to the fears of some observers). For the most part, the trees that seeded in were similar to what had been there before, seeding in from nearby surviving trees or the lodgepole's serotinous cones. The exception present in many burned areas was aspen seedlings, a surprising discovery because the tree had rarely been observed to germinate from its own seeds, depending instead on clonal disbursement (sprouting from the roots of nearby trees). Even more surprising was that some of the seedlings were miles from the closest aspen tree capable of producing seeds. Despite being heavily browsed by elk, aspen seedlings/saplings were still abundant in 2010, some growing above browse height, especially in the higher elevations and in places where they were protected by tangles of fallen fire-killed trees. Last, the elk themselves, one of Yellowstone's most numerous and frequently sighted large animals, were still another research subject. Between the burned forage and the severe winter after the fires, the elk population took a nosedive, declining 38–43 percent by spring 1989. Indicative of a population under stress, calf birth weights were down 17 percent that calving season, and calf mortality was twice the typical rate. However, if the fires were a short-term disadvantage, they were more of a long-term bonanza, as the nutrients locked up in dead plant materials were made available to the plants that survived or germinated, and then to the animals that ate those plants. For example, while elk avoided burned grasslands the winter following the fires, they sought them out 40–50 percent more often than unburned grasslands two to three years after the fires. By 1995, Yellowstone's elk populations were back at their prefire numbers, and one more fascinating piece of the Yellowstone ecosystem was illuminated.[11]

As the years after the fires became decades, however, our view of Yellowstone's fire story began to change. Yellowstone's megafires came to have company, with more and more western states experiencing massive fires, often with hundreds or thousands of homes destroyed. At first, most of us blamed the decades of fire suppression that many, if not most, of those forests had seen, leading to higher-than-normal fuel loads and undoubtedly contributing to the fires' increasing frequency and ferocity. Others pointed to the persistent drought that gripped the West in the 2000s. Indeed, the first decade of the new millennium saw a significant increase in both the number and size of large fires; for example, there were nine times as many large fires in the West between 2003 and 2012 as there were in a comparable decade before the ramped-up fire era, 1973 to 1982.[12] The 2010s saw the fires grow more, with five of the eleven western states (California, New Mexico, Arizona, Oregon, and Washington) all seeing their largest fires in history. The six other western states either saw their largest fires the previous decade (Utah, Nevada, and Colorado) or had unusual mitigating circumstances (Montana and Idaho fires are still dwarfed by the Big Burn—even though Idaho had a 653,000-acre fire in 2007—and Wyoming had the 1988 Yellowstone fires). Since the year 2000, then, every western state experienced its largest fire in history, with the exceptions of the states in which the Big Burn and Yellowstone fires occurred, all of which, nonetheless, had massive, relevant fires. With one exception, these megafires have all been well over 200,000 acres in size, with four of them exceeding 500,000 acres. Some have been terribly destructive, such as the 2018 Camp Fire, which burned 149,000 acres and most of the town of Paradise, California, including 18,800 structures, most of them homes. The fire also killed eighty-six people, making it the deadliest and most destructive fire in state history.[13] Yellowstone's 1988 fires, then, were the first in a lengthening list of increasingly large, deadly, and damaging fires in the West.

Size and notoriety are not the only things that Yellowstone's signature fires have in common with the subsequent megafires, which collectively had less than a 2 percent chance of happening due to random variation.[14] Like Yellowstone (whose climate changes are presented in the next paragraph), snowpacks are declining throughout the West, so much that 90 percent of long-term snow-monitoring stations are recording declines. A third of these are significant declines, and they are happening in all states, months, and climates, but especially in spring and in areas with milder winter climates, like the West Coast states. Since the mid-1900s, between 15 and 30 percent of the region's snowpack has disappeared, a snow water equivalent amount equal to the capacity of Lake Mead, the nation's largest reservoir.[15] Rising temperatures in February and March are driving the snowpack loss, both through conversion of snow to rain and also through earlier melting, with rain and temperatures above freezing.[16] In turn, the warmer temperatures and earlier spring melt-off are the primary cause of the recent surge in large western wildfires.[17] Climate models indicate that between 20 and 50 percent of the West's snowpack loss is due to natural climate cycles, while the rest is human caused.[18] In fact, almost half the acreage burned since 1979 can be attributed to human actions—an additional 10.4 million acres gone up in smoke—vis-à-vis increasing the aridity of wildfire fuels and adding nine days annually of high fire potential.[19] The overall fire season throughout the West is now seventy-eight days longer than it was before 1986.[20] Summing up, human activity is exacerbating natural climate cycles in the West, shrinking snowpacks, melting them out earlier, drying wildfire fuels, and ultimately producing more, bigger, deadlier, and highly devastating wildfires.

The situation is little different in Yellowstone, except for the fact that fewer people have died or lost their homes—yet. Compared to its historic climate, Yellowstone's contemporary climate is at the extreme hot end by almost any measure; the annual average temperature, the July average high temperature, the January average low temperature, the winter average temperature, and the summer average temperature are all essentially in record terrain.[21] More specifically, temperatures throughout the Yellowstone region have been increasing since 1948, at about 0.31 degrees per decade, with the warming somewhat skewed toward the northern end of the Greater Yellowstone Ecosystem. The park itself is warming at 0.29 degrees per decade, totaling about 2 degrees since 1948.[22] As expected, higher temperatures, especially in winter and spring, are shrinking snowpacks throughout the Yellowstone ecosystem; of thirty snowpack-monitoring sites, 70 percent of them (twenty-one sites) show significant snowpack declines since 1961. The snowpacks of the early 2000s are markedly low, so minimal that they rival those of the Dust Bowl era; in fact, tree-ring analysis shows that both are the lowest in eight hundred years. Demonstrating the sensitivity of the climate system is the fact that March temperatures at the declining snow stations have only increased 2 or 3 degrees, but that has been enough, when combined with winter warming, to push days above freezing at the northeast entrance from eighty to a hundred. There, winters are a full month shorter than in the 1960s, with snowpacks just over half what they were then. The thinner snowpack takes less time to melt off (thirty days now, down from forty-six), and the typical melt-out day has moved from May 29 to May 5. With most other long-term snowpack-monitoring sites showing the same decline, it's clear that, throughout the Yellowstone ecosystem, winter and spring are getting warmer, less snow is falling and accumulating, and that snow is melting off earlier. The consequences for the vegetation of Yellowstone, as we have already seen, have been severe—and we will soon see other trees that are not faring well in the hotter, drier climate we are creating.[23]

It is largely our actions driving Yellowstone's warming temperatures and diminishing snowpack, just as they are throughout the West. In Yellowstone, we're responsible for at least half of the warming increase and as much as 98 percent, depending on the specific snowpack-monitoring site.[24] In the park itself, I suspect the human contribution has been at the low end of this range, given that many Yellowstone forests were senescent, with plenty of fuel to sustain a fire. In the end, the 1988 fires were a product of both nature's forces and of our own; where I once saw the omnipotent power of nature, now I see hints of human-caused destruction. The influences of nature are still prominent, probably more evident than in most western landscapes, but our negative influence is there as well: anthropogenic global warming has helped fuel the megafires throughout the West in the last twenty years, making climate change the common denominator in these gigantic blazes. Nature's cycles were probably a factor in all these infernos, but the influence of climate change is growing, and will soon be the dominant influence in the megafires—if it isn't already.[25]

As I settled in with my parents in St. Louis eight years after my hike with Josh, I initially found life with them to be easier than living alone in Gardiner (even with the assistance of my friends). I had brought my trike with me, and enjoyed riding it around the lake in their neighborhood. They had obtained a loaner motorized wheelchair, which I used anytime I went outside, reducing the falls I'd been suffering in Montana. They had also remodeled their ground floor bathroom to make it accessible, removing the risks involved with showering at my Gardiner house. Even with the grab bars and handrails we had installed in my Gardiner bathtub, stepping out of it was becoming risky for me. In fact, the week before I moved to St. Louis, I had fallen doing that move, ending up in the awkward position of lying crossways across the tub, with my head against the back wall and my knees over the edge of the tub. My buddy Jim Roche from Yosemite was visiting then and heard my head hit the shower wall, so he came running when I called for him, unable to get up.

However, he couldn't pull me up and out by himself, so we called another friend to help. The two of them succeeded, and I was more embarrassed than injured (at exposing myself to them, if accidentally), but it was another illustration that I was nearing the end of living independently, even with assistance. Once resettled in St. Louis, I was able to shower without risk of falling, for the shower was barrier-free with handrails in better positions. It would not be long, though, before even those were not enough for me to bathe myself, and the disease progression returned to making life more difficult and degrading.

The first indication of continued progression came when I was biking at a nearby state park. The autumn day was sunny and warm, so I tried to pull my T-shirt off to enjoy the warmth on my skin. Try as I might, I could not pull the shirt up and over my head. Just a few weeks earlier—really, just a few days—I had been able to make that move. That experience set the pattern for the fall, as my arm and hand strength rapidly eroded. By Christmas, I was struggling to lift my hands up high enough to shampoo my hair. I fought against asking for help from my dad for a few days, tilting my head to the side and washing my hair in halves, but that didn't really work. Some ALS sufferers have likened the progressing muscle atrophy to having ten-pound weights on your hands with a pound added every day, which is a good comparison. With no way to remove the now-thirty-pound weights from my hands, I asked Dad for help. Since then, he has dutifully bathed me every other day (not being active enough to break a sweat, daily showers became unnecessary). In a few more months, I witnessed my arm strength ebb another way, this time at the kitchen table. When I sat down there, I would lift my hands up to rest on the table, as we all do without thinking. As my arm strength eroded, it became increasingly difficult to lift my hands up enough, and I could actually track the rate of decline by measuring the difference between the height I achieved compared to the tabletop. It became a game of desperation and futility, seeing my strength give way so quickly—an inch a day sometimes—with the end result being that my hands stayed on my lap till someone lifted them for me.

My hands, meanwhile, were also weakening, so I used a fork and spoon with extra-large handles that I grasped with my whole hand, not just two or three fingers as we ordinarily do. As you would guess, the special utensils became pointless once I couldn't lift my hands to my mouth, whereupon my folks fed me. You can imagine how that makes a person feel, being fed like a toddler. Using the keyboard, though, was where the hand and finger weakening was more obvious. I was able to type regularly when I left Yellowstone in October, and was still typing normally in January. That began to change as the winter wore on; by March, it felt like my fingers were dipped in honey, for they lingered on the keys, giving me strings of the same letter. I would spend as much time correcting a sentence as I did typing it in the first place. A half-page letter to my doctor took two hours to write and correct, and progress on the book dropped to a crawl. Increasingly frustrated, I bought a program that tried to predict words as I typed, but it was little help, designed for users with other communication problems. I could still click the mouse strongly, so I activated Microsoft's on-screen keyboard and improved my "typing" speed modestly. This solution left a lot to be desired—I knew there was better word-prediction software, but I couldn't find a compatible and affordable version. However, it was preferable to sticky-finger typing, so I soldiered onward and had a manuscript ready to submit to a publisher by springtime.

Completing the manuscript felt good, a welcome victory in a losing war, a momentary distraction as I watched the last remaining vestiges of independence slipping away. No longer could I type, feed myself, or shower, and the loss of hand and arm strength made many other life activities difficult or impossible. Driving, the most tangible symbol of personal independence for most of us, disappeared before I had even left Yellowstone, its risks too great to ignore. Biking became the next casualty, my arms too weak to lift my hands to the handlebars on my trike, and my fingers too weak to work the gearshift. The list of casualties became too long to count as the myriad things we do with our hands became impossible, everything from swatting a mosquito, scratching an itch, holding a telephone, cleaning my glasses, writing a check, dressing myself, opening a door, and turning the page to shaking hands and caressing a loved one. Indeed, by the time I submitted the manuscript to my editor, I had long passed the point of needing full-time care.

This list of woes was still in my future when Josh and I made camp at the base of Eagle Peak on our third day in Yellowstone's Thorofare. We had left the Yellowstone River Valley after a few miles, going up Mountain Creek and then its tributary, Howell Creek. With most of the Mountain Creek drainage burned in 1988 and with many meadows along the two creeks, the valley was mostly open, giving us good views of the Absaroka Mountains ahead. The Absarokas are a long chain, stretching from fifty miles north of Yellowstone to fifty miles southeast of the park, forming the park's entire eastern boundary, where the range is at its widest, about seventy-five miles. Within Yellowstone, the Absarokas top out at 11,372 feet (Eagle Peak), but the range slopes up to the east, culminating at 13,153 feet above sea level at Francs Peak.[26] The range was formed over ten million years (beginning about fifty-three million years ago) as many different volcanoes and vents erupted, including some vents in the Eagle Peak area. The ash, pumice, and pyroclastic flows (slurries of hot volcanic debris resembling liquid concrete) accumulated to depths of several thousand feet, with the total volume estimated at about twenty thousand cubic miles. Much of this material was hot enough to weld loosely together once on the ground, producing an easily eroded, cobbled, fragmented rock that one of my hiking partners called "Absaroka crumble."[27]

Erosive or not, the volcanic field has endured forty-three million years of erosion and glaciations, lowering the mountaintops and carving the valleys. Absaroka crumble may not be strong enough to create the cliffs and valleys of Yosemite, but the soils derived from it are rich in the minerals plants need, like potassium, magnesium, phosphorus, and calcium. Absaroka crumble is low, though, in precious metals, so its mountains never saw the prospecting and mining activity that other mountains saw, or still see. Last, Absaroka crumble does not make safe rock climbing, nor does it make many lakes (glaciers left just a few, and many of those have been converted to meadows through sediment deposition). There are exceptions to all these generalizations, but the net result is a land of rich plant and animal diversity and fecundity (herbivores and carnivores), few roads (leading to the mines that were never there), and little recreational use as compared to other wilderness areas that have climber and hiker magnets (cliffs of strong rock, and lakes). In other words, the Thorofare's size and wildness are due in part to its underlying geology, and the people who do venture deeply into it are likely to have several days without seeing other people. Unlike other wilderness areas, especially in California but even in Wyoming, the Thorofare's wilderness areas (the Washakie and Teton Wildernesses, along with Yellowstone's southeast corner, managed as wilderness) live up to their name as wilderness, and all because of Absaroka crumble (well, not all, but more than just a little).

The fourth day dawned clear and cool for Josh and me, just 2 degrees above freezing—perfect weather for a summit attempt. We began the adventure by bushwhacking three miles up the valley of the unnamed creek we'd camped on, an unnamed tributary of Howell Creek. The valley was open, with several meadows and with most of the 1988 fire-killed trees still standing, not on the ground where they can make deadfall tangles—sometimes three or four layers of downed trees—that force one to detour around or climb up over. As we hiked up-valley, we passed below Table Mountain, its east face an eight-hundred-foot vertical cliff above and a talus slope of equal height below. The cliffs are banded with different shades of gray, the bands being the various layers of ash and pumice thrown out by volcanoes millions of years ago. I had climbed Table a decade before, using a route that did not involve scaling cliffs, and finding a plateau of treeless tundra on top, awash in the colors of the summertime alpine-flower bloom. Looking across at Eagle Peak then, I had seen a route to its summit that also did not involve scaling any cliffs, as the more popular summit route on the other side of Eagle did (two short cliff bands, but potential showstoppers nonetheless). I filed that knowledge away in my hiking memory, but am still surprised, looking back now, that it took a decade to get back there.

When Josh and I arrived at the head of the valley (the ridge forming Yellowstone's east boundary), though, I initially doubted my recollection. We had only five hundred feet of climbing left to go (which would make three thousand total for the day), but what we first saw on Eagle Peak was a mountainside of steep, rocky slopes that offered only perilous, sliding footing crowned with a foreboding ring of jagged, fragmented cliffs.

Studying it more, though, I saw that we could follow the ridge we were on, then sidestep around the toe of the summit crown to what appeared to be a couloir (a chute through impenetrable rock layers), and follow it up to the mountaintop. That route was indeed passable, leading through the dangers—and through the rock layers, which all seemed tilted to the north, as though the entire mountaintop had leaned in that direction. We did not notice that at the time, but it's strikingly evident in the photos I took that day. What we did notice was the mountain's rockiness; that is, nowhere on the mountain's sides or summit was there any tundra—in fact, but for a few isolated clumps of krummholz, the mountain was completely naked, devoid of soil and vegetation. This may be typical of other mountain ranges, but it's less common for the Absarokas, which have so much tundra that the dominant color of some Absaroka peaks in summertime is vibrant green. The rockiness and slant of Eagle Peak could be related; perhaps upwelling magma in the nearby volcanic vent forcefully bent the rock layers upward and northward, leaving the layers fragmented and shattered, producing both the mountain's slant and its rockiness. Whatever the cause(s), the mountain's secrets remain untold, part of the magic and mystery of Yellowstone and wilderness.[28]

Once we found comfortable rocks to sit on (Eagle Peak's summit is as cobbled as its mountainsides), we stopped for lunch with an unbeatable view. In every direction, mountains sprawled, valleys yawned, and space consumed all. To the west, the Tetons punctuated the horizon, while the Yellowstone River meandered through the trappers' thoroughfare far below, the flat valley giving way as the eyes moved north to the arms of Yellowstone Lake. In the other directions, mountains piled upon mountains, their cliffs and slopes varying in color from dark purple to chocolate and back again, the tundra among them already the rusts of autumn. Their shapes took almost every conceivable mountain form, while their abundance stretched to the limits of our sight. Seemingly providing an opposite version of mountain shape and form from our mountaintop was Mount Humphreys, the next peak north on the boundary ridge (see plate 21). Its summit is flat, large, and vegetated, and its sloping sides are striped with stringers of trees alternating with avalanche chutes, the trees shrinking in height as their growing season shrinks in length nearer the

PLATE 1 Hoh River Valley, Olympic National Park, August 22, 2013. The Hoh receives twelve to fourteen feet of rain each year, making it the wettest place in the continental United States. Photo from Michael J. Yochim photo collection.

PLATE 2 White Glacier and Glacier Creek, Olympic National Park, August 20, 2013. Olympic has nine glaciers, containing 80 percent of the park's ice. Photo from Michael J. Yochim photo collection.

PLATE 3 Sunrise on Blue Glacier and Mount Olympus, Olympic National Park, August 21, 2013. Like most glaciers worldwide, Olympic's glaciers are retreating. Photo from Michael J. Yochim photo collection.

PLATE 4 Blue Glacier from the terminal moraine, Olympic National Park, August 21, 2013. Since 1986, Blue Glacier has retreated fifty feet per year. Photo from Michael J. Yochim photo collection.

PLATE 5 Lupines and bear grass near Seven Lakes Basin, Olympic National Park, August 22, 2013. Photo from Michael J. Yochim photo collection.

PLATE 6 Heart Lake, Olympic National Park, August 23, 2013. The three shades of green covering the land added to the subalpine beauty at this lake. Photo from Michael J. Yochim photo collection.

PLATE 7 Grand Canyon National Park, Arizona, May 2014. The canyon plunges 1 mile deep, opens up to 18 miles wide, and flows 277 river miles in length. More than a third of Earth's geologic past is held in its layers. Photo by Paul Yochim.

PLATE 8 Rafts on the Colorado River, Grand Canyon, May 2014. Photo from Michael J. Yochim photo collection.

PLATE 9 *Left to right*: Paul, Mike, and Jim Yochim, Phantom Ranch, Grand Canyon, May 2014. Photo by Bob Graves.

PLATE 10 Colorado River at Nankoweap, Grand Canyon National Park, May 2014. The desert pictured here will see its average daily July high temperature move from an already torrid 100 degrees to a scorching 112 degrees by the year 2100, if present trends continue. Photo from Michael J. Yochim photo collection.

PLATE 11 Mike Yochim at Deer Creek Falls, Grand Canyon, May 2014. Hiking poles were essential at this stage of the disease. Photo by Paul Yochim.

PLATE 12 A near-perfect day on the summit of Boulder Peak, Glacier National Park, August 2007. Mike's wilderness companion, Woody (*on the right*), succumbed to cancer less than a year later. Boulder Peak had a glacier—with an ice cave—on its north side when the park was established in 1910, but there is nothing left of either now. Photo from Michael J. Yochim photo collection.

PLATE 13 Hole in the Wall, Glacier National Park, August 4, 2007. The valley in the left foreground is a "hanging valley," perched above the valley to the right and below. Photo from Michael J. Yochim photo collection.

PLATE 14 Last light from Hole in the Wall camp, Glacier National Park, August 4, 2007. Photo from Michael J. Yochim photo collection.

PLATE 15 Boulder Peak, Glacier National Park, August 2007. The author climbed it on a day bordering on perfection, a day where the senses were in overdrive. Photo from Michael J. Yochim photo collection.

PLATE 16 Fireweed at Boulder Pass, Glacier National Park, August 4, 2007. Photo from Michael J. Yochim photo collection.

PLATE 17 *Left to right*: Mount Peabody, Pocket Lake, Agassiz Glacier, and Kintla and Kinnerly Peak, Glacier National Park, August 4, 2007. Photo from Michael J. Yochim photo collection.

PLATE 18 Goat Haunt sunrise, Waterton-Glacier International Peace Park, July 31, 2014. The Canadian and American governments coordinate management of Waterton-Glacier. Photo from Michael J. Yochim photo collection.

PLATE 19 Sunset on upper Mountain Creek, Thorofare, Yellowstone National Park. August 30, 2006. The Thorofare is the most remote spot in the continental United States. Photo from Michael J. Yochim photo collection.

PLATE 20 Mike Yochim near Mount Holmes, Yellowstone National Park, June 10, 2007. A rare photo where the author is not wearing sunglasses or a hiking hat. Photo by Ellen Petrick.

PLATE 21 Mount Humphreys, the next peak north on the boundary ridge from Eagle Peak, Yellowstone National Park, August 29, 2006. Photo from Michael J. Yochim photo collection.

PLATE 22 Southeast Arm ("as blue as the Rocky Mountain sky"), Lake Yellowstone, Yellowstone National Park, August 26, 2006. Photo from Michael J. Yochim photo collection.

PLATE 23 The author on the summit of Eagle Peak, the highest point in Yellowstone National Park, August 29, 2006. Photo from Michael J. Yochim photo collection.

PLATE 24 A twice-burned hillside below the Trident in the Thorofare, Yellowstone National Park, 2006. Since 2000, Yellowstone has experienced several such "reburns," a fire behavior not previously seen there. As is evident by the lack of seedlings, some such reburns may not return to forest—a pattern that observers fear could become widespread throughout the region by midcentury. Photo from Michael J. Yochim photo collection.

PLATE 25 Mike Yochim in Yosemite Valley, with Yosemite Falls in background, 2012. He worked at Yosemite for five years as a senior planner for the NPS. Photo by Jeanne Yochim.

PLATE 26 *Left to right*: Mike and Brian Yochim at Tunnel View, Yosemite National Park, November 6, 2011. Photo by Jill Yochim.

PLATE 27. View from Parsons Memorial Lodge, Tuolumne Meadows, Yosemite National Park, July 1, 2020. Parsons was built by the Sierra Club in 1915. Mike skied to it and found it buried under three to four feet of snow. Photo by Brian Yochim.

PLATE 28 Snow surveyors arrive at a backcountry cabin, Yosemite National Park, February 2012. Surveys like this one are done monthly throughout the Sierra winter to forecast the amount of snowmelt available for irrigation and municipal water providers. They also record a long-term decline in snowpack throughout the state. Note the general lack of snow. Photo from Michael J. Yochim photo collection.

PLATE 29 Cathedral Peak (*on the left*) and Tuolumne Meadows, from Lembert Dome. This area was once buried under a glacier two thousand feet thick. Photo from Michael J. Yochim photo collection.

PLATE 30 Headwaters of the Tuolumne River: Lyell Canyon and Glacier from the air, April 2011. Photo from Michael J. Yochim photo collection.

PLATE 31 Mike Yochim on the Trans-Sierra ski trip, with Tenaya Lake in the background, April 17, 2011. Photo from Michael J. Yochim photo collection.

PLATE 32. Mike Yochim working at his desk in St. Louis, Missouri, January 5, 2020. Photo by James J. Yochim.

summit. Many mountainsides in view had similar vertical striping, the trees a mix of burned and not. Clouds and meadows provided contrast in the landscape, the one dancing across the landscape with shadows in tow, the other adorning Howell Creek like strings of bright emerald. Another contrast was as much felt as seen: the day's warmth, which had us both in shorts and T-shirts. The balmy weather was defied by a snowfield tucked into a sheltered alcove on the mountain's north toe. Being steeply pitched with crevasses on one side, the ice patch could have been a glacier in times past, but is not big enough (at least twenty-five acres) to be a glacier today. Still, it was a reminder of the massive glaciers that carved out the valleys all around us, rivers of ice that left their mark on the landscape in the form of U-shaped valley cross sections (valleys carved by water alone usually have a V-shaped profile). Ice patches, meadow, mountains, sky, forests, and grandeur all were gifts we gazed upon for well over an hour that awesome day, a time I will long remember.

With nothing human anywhere in view, it felt like we could have been the first humans to ever dwell upon that scene.[29] Certainly, that's how it felt at the time, even though we know that Native Americans have been present throughout the Yellowstone region for more than ten thousand years, especially in the Yellowstone Lake area in the warmer seasons.[30] But in the years since Josh and I sat on Yellowstone's highest peak, we've come to understand that our society is having an increasing role in shaping the forests and landscape of the Yellowstone ecosystem. To begin with, although Yellowstone contains no glaciers, glaciers in the surrounding mountains—the Teton, Wind River, and Beartooth mountain ranges—are all retreating.[31] The Absarokas have but two glaciers, the larger of which, Fishhawk Glacier, is just a few miles east of Eagle Peak. Little studied or photographed, the glacier is probably retreating, given the retreat that all other Wyoming glaciers are experiencing, though I was unable to confirm that assumption.[32] Also, many of the forested valleys flowing away from our perch atop Eagle Peak had burned in 1988, so the global warming / human influence was certainly present. Another vegetative change indicative of a warming climate is in the aspens that germinated after the 1988 fires. In 1999, the aspens at lower elevations were more favored, but by 2013, those at higher elevations were doing better. Declining elk populations may be a confounding factor in this upward shift, but increasing temperatures are the probable cause.[33] Similarly, the 1988 fires catalyzed a shift upslope in the lower treeline, especially in areas with high fire intensity.[34] Plants at the lower treeline are experiencing a long-term reduction in the groundwater available to them, with perpetual drought conditions being the norm since 1997.[35] Finally, by 2006, most of the tundra in the higher Absaroka Mountains and Wind River Range no longer met a commonly used classification of vegetation types, the tundra temperatures having risen above the 50 degree July average temperature threshold.[36] All these findings are consistent with a warming climate causing plants to shift upslope as lower elevations become too dry and hot for them. While Josh and I could not have seen all of these changes even if we had known of them in 2006, we could definitely have seen many of them—and seen that nature's forces were increasingly giving way to ours.

Perhaps the most ecologically significant change was still to come to the Yellowstone area, though its foundations were being laid the year we did this hike. Among the trees we could see from the summit, such as in the stringers climbing the nearby peaks, were whitebark pines. Whitebarks live high on the mountainsides of greater Yellowstone, often up to the upper treeline. They are not a tall tree—fifty feet is exceptional—but with an open crown, they are an especially lovely component of those rarefied forests, otherwise dominated by the spires of spruces and firs. Their cones, rich in fat and protein, were a mainstay in the diet of grizzly bears fattening up for hibernation. Red squirrels, common in Yellowstone, relied on the nuts as well, cutting the cones from the treetops, collecting them into hoards known as middens, and then retrieving the cones throughout the winter to eat the nuts inside. Because the cones were indehiscent (they don't open on their own) and remained on the tree's uppermost branches, hungry grizzlies had to find and raid squirrel middens. The rich nuts are also the tree's dispersal mechanism, with Clark's nutcrackers (a bird in the jay family) the agent, collecting and caching thousands of nuts, some of them buried and others hidden aboveground, where they can be retrieved when snow covers the ground. Invariably, the birds forget some of the buried nuts, despite their amazing spatial memory (thought to be one of animal kingdom's greatest). Nutcrackers have been observed flying up to twenty miles to forage, ensuring the tree's dispersal ability. For these reasons, the tree is considered to be a keystone species, as much a linchpin in subalpine forests as a keystone is to an arch.[37]

Whitebark pines are under attack in the Yellowstone ecosystem from at least three causes in addition to drought. The most significant of these is the mountain pine beetle, a native insect that lives in the growing tissues of pine trees. The adult females emerge from the tree in which they were born, fly to a new, live tree (usually within three miles but farther in favorable winds), attract mates, tunnel into the phloem, and lay their eggs. The larvae feed on the phloem, pupate, emerge, and complete the life cycle while girdling the tree.[38] The onset of warmer, drier weather in the 2000s has affected whitebarks by increasing the probability of mortality when attacked by the beetles. Irruptions of the beetle have occurred in the past, but global warming exacerbated their threat in at least three ways: First, with a lengthening growing season and warmer fall temperatures, the beetle is able to complete its cycle in one year, not two, effectively doubling its response time and its numbers. Second, with warming winters where temperatures no longer fall below -40 degrees—the temperature at which beetles are killed—still more beetles survive. And third, like most trees, whitebarks are stressed by drought, reducing their ability to fend off beetles, which they ordinarily do by flooding their tunnels with sap.[39] The years from 2006 to 2008 brought warmer-than-normal fall and winter temperatures and reduced precipitation, creating the perfect storm for the mountain pine beetle and its unfortunate hosts. The resulting beetle epidemic peaked in 2009 and resulted in 79 percent of the whitebarks in the Yellowstone ecosystem dying.[40] It was the worst beetle outbreak in history, with sixteen of the twenty-two major mountain ranges in the Yellowstone ecosystem sustaining moderate to severe whitebark mortality. The sites where the pines persist in the other six ranges are those with the coldest microclimates.[41] Such sites may be the only places where whitebarks occur by the end of the century if present trends continue, according to a recent study that used nine global climate models to project the tree's distribution. The study authors noted that in all nine models, the pine's range began to shrink when the annual average temperature rose

1.8 degrees from the historic average.[42] With Yellowstone's annual average already up by 2 degrees, and given the changes the beetle outbreak has already caused, it's likely that the range contraction is already underway.

As if the beetles were not enough for the signature tree to contend with, white pine blister rust and changing fire regimes are also challenging the remaining whitebark pines. Originally brought to America with European nursery stock in the early 1900s, blister rust infects all age classes of whitebark pine, killing most trees within four years of infection. Present in the greater Yellowstone region since 1937, the rust is now widespread throughout the Yellowstone ecosystem.[43] Finally, whitebark faces the changing fire regimes in subalpine forests. In the past, whitebark had a competitive advantage over the trees with whom it shares the high elevations, subalpine fir and Engelmann spruce: it could survive low-intensity fires, while its competitors could not. In contrast with the other two (especially the fir), whitebarks have thicker bark and a more open crown and few low branches that fire can use as a ladder to climb into the canopy. Now, fires are more intense, traveling more into the whitebark forests and killing them. In this way, fire as friend has morphed into fire as foe, compounding the losses to beetle and rust.[44] Not surprisingly, the US Fish and Wildlife Service recommended that whitebark pine be designated a threatened species under the Endangered Species Act in 2011.[45] Clark's nutcrackers, while far from being recommended to the endangered species list, are displaced by the wholesale die-off of whitebark pine that has occurred in the Yellowstone region. The bird is an example of the ripple effects that climate change is causing.[46]

Back on Eagle Peak, Josh and I prepared to leave the summit and savored the view for one final moment. It was a landscape we thought was wholly wild, dominated by nature's forces. The burned forests were perhaps the most visible evidence of that wildness, especially to those of us who had seen the fires march across the Yellowstone landscape in 1988, in spite of our herculean efforts to corral and control them. Now, though, we see the hand of humanity not just in the forests burned in the 1988 fires, but also in the ghostly skeletons of whitebark pines, the warming tundra, and the rising lower treeline. Everywhere we look, we can see the signature of climate change, its destructive influence widely discernible to the informed observer. There is more than a little irony in realizing that, in restoring the natural force of wildland fire to the Yellowstone landscape, park managers inadvertently created a western landscape that displays the hand of humanity vis-à-vis climate change more than most other less studied, western landscapes. They could not have known at the time that climate change was enhancing the potency of wildfire, but the evidence is clear now: climate change is here, and it's growing in size and power. Through no fault of their own—yet everyone's fault—park managers have created a landscape of climate-altered forests.

In 2014 life in Missouri, the muscles in my head and throat were weakening as fast as those in my arms and hands, which is characteristic of the bulbar form of the disease. When I first arrived in St. Louis, my speech was obviously slurred, but most people could make out what I was trying to communicate. For more detailed communications, I would type an email to my folks, friends, and colleagues. The passage of time made my speech increasingly thick, slurred beyond recognition even at times to my folks, who knew my speech patterns better than anyone. It was increasingly clear that we would need some help soon, so we saw a speech pathologist, who introduced us to the world of assistive and augmentative tools. In my case, these were speech devices: computers that have both hardware attachments that track my eyes moving about an on-screen keyboard, and also software that translates the typed message into speech. Initially, I just used the device to converse with visitors, but as my speech eroded, I came to rely on it more and more. Speech devices are the cat's meow of assistive tools for ALS sufferers, even allowing me to write this book and *Essential Yellowstone*. Truly, I can't imagine what it must have been like for earlier ALS victims without such devices; I doubt I would still be alive without mine enabling me to stay engaged in creative, productive work. As good as they are, though, they have limitations. For example, having to type what one wants to say is nowhere close to the speed of typical conversation (especially when typing using an eye-tracking device). I find myself just listening to most conversations, having to begin my responses with brief reminders of what question I'm answering—five or ten minutes later—and missing parts of the conversation while I focus on composing my response. For that reason, the speech device's ability to keep ALS victims as fully engaged members of society is limited. Without speech, it's impossible for ALS sufferers to remain in the role they once filled, despite technological advances.

Like anything in the American health-care system, the devices don't come cheap; current models start in the $14,000 range, and most insurance plans don't cover them. I was able to borrow an older model from a woman who no longer needed hers, and eventually to upgrade to newer versions through the Team Gleason Foundation, an organization founded by former football star Steve Gleason after he was diagnosed with ALS. The group lends speech devices and other assistive devices to ALS sufferers. For most of us, however, the first high-ticket item needed is a motorized wheelchair, which costs as much as a good car. Mine was $24,000, of which I paid about $5,000 (insurance paid the remainder). Wheelchairs alone don't get a person to the grocery store or doctor's office, so many ALS patients need to buy a vehicle with a wheelchair ramp, so there goes thousands more dollars. Remodeling one's house and bathroom to make them accessible is another expense that many disease victims incur, taking another chunk of savings. Treatments to slow the disease progression are next—there are but two of them, both of them expensive, neither of them spectacularly effective. The first of these, an oral medication called riluzole, has been available for a few decades and costs $24,000 per year; it extends a patient's life by two or three months. The second treatment, an intravenous drug called edaravone, costs $146,000 annually; it slows the disease progression by a third. Questions persist about the efficacy and value of both treatments, but most insurers do cover them. Perhaps the single most costly item for most ALS patients, however, is the care required for them, especially those in the advanced stages of the disease. Akin to quadriplegics, they are unable to care for themselves at all, needing full-time care. For parts of the daily routine such as dressing, two caregivers are usually necessary. The disease most commonly strikes between ages forty and sixty, which is when many people are earning their highest lifetime incomes. Not only do those with

ALS have to quit working, but their spouses often need to do so as well, to provide care. Households afflicted by ALS, then, can see their income stream shrink significantly or disappear altogether, right when they need it most. While ALS sufferers are eligible for Social Security Disability Insurance payments and, depending on their state of residence, some state assistance, those programs together usually provide just a fraction of the family's predisease income. In these and other ways, ALS is as much a financial challenge as it is a physical one.

Returning to some of my physical challenges, swallowing safely became an increasing problem as my time in Missouri stretched past a year. The muscles in my mouth and throat were weakening, and my epiglottis was slow to close, causing me to choke regularly. Liquids and crunchy foods like crackers and chips were particularly risky for me, as my weakened tongue struggled to keep the droplets and pieces corralled. Munching on Wheat Thins one day at lunch, for example, a piece of cracker too big to swallow slipped into my throat, getting stuck crossways. I could still breathe freely, but an attempt to swallow it resulted in it getting more stuck, the cracker poking painfully in the opposing sides of my throat. Not knowing what else to do, I hoped that it would soften in my throat's humid environment, which it did, in only a couple of minutes. Taking a deep breath, I swallowed hard, and it went down, smoothly—and that was the last cracker I ate. Occurring much more frequently were larynx spasms, in which the larynx muscles move tightly together, closing the windpipe. Usually triggered by food particles, liquid, or phlegm getting past the epiglottis, such spasms may be the body's last defense against foreign substances invading the lungs—but with no air entering the lungs either, these episodes are quite frightening. After a few seconds, though, the spasm passes, more quickly if I remain calm and try to inhale through my nose. Those are both hard things to do when you're suffocating, and sometimes the spasms don't release quickly—or they occur in multiples—so they remain frightful, more than five years after experiencing my first one. These were the more intense indications that my ability to swallow safely was eroding—more common ones were sticky foods like peanut butter getting stuck to my palate, my tongue too weak to push it off; the increasing time it took for me to finish my plate; and my growing reliance on drinking Ensure to compensate for the dwindling caloric intake. I also went out to eat less and less often, not sure if I would be able to find anything on the menu I could eat safely. Eating was losing its appeal, turning meals into a chore, often lasting an hour or more and sometimes leaving me exhausted.

Complementing my face and throat problems was my neck, its muscles weakening so much that they could no longer support my head. I had arrived in St. Louis with this problem, but it continued to get worse, so I spent increasing amounts of time with a brace wrapped around my neck. One particular incident highlighted the mounting difficulties I was facing at the time, the increasing demands on my caregiver parents, and the awkward situations the disease can create—and the growing isolation being forced upon me. We had visitors for an hour or two, and both my parents were out of the room, preparing drinks. Somehow, my head slipped out of the neck brace, falling down to the side. My neck was too weak to lift it up, I could not tell the visitors what to do intelligibly, and they were too unfamiliar with the disease and its predicaments to recognize that they needed to do something. So there we sat, looking at each other at an awkward angle and waiting for my parents to return. They came back a few minutes later and promptly lifted my head to a normal position. The incident has lingered in my memory, perhaps because it epitomized the frustrations ALS forces upon those it strikes, from the victims themselves to their families and caregivers and even beyond, to casual acquaintances and neighbors. Moreover, between the speech challenges and the eating and swallowing problems, the affliction threatens the most universal of social venues: breaking bread together and sharing a meal. Indeed, if I could have one ability back, my first choice would be speech—and the second would be the ability to eat safely.

The morning after we'd climbed Eagle Peak, Josh and I broke camp and embarked on a leisurely, scenic day, beginning with an easy hike to a nearby meadow, outside the park. Actually a series of openings, the meadows along Mountain Creek are separated by lines of live trees, some of them thin enough that we could see the next meadow through them. The last meadow was a place of exceptional mountain beauty, an almost-flat field of sun-cured, flaxen grasses nearly encircled by Absaroka peaks, with names evoking the yin and yang of mountains reaching for the sky: Chaos Mountain on the north (left), and Overlook Mountain on the east (in front of us). Linking the two was a lofty mountain ridge that wrapped around the meadow, forming its southern wall and continuing on to the Trident, overlooking the Thorofare several miles to the southwest. All these mountains and ridges exceed eleven thousand feet, more than a vertical half mile above the meadow. The meadow was a mountain hideaway, a sheltered fastness seen by only a few—and it was unoccupied, so we found a place to pitch the tent and set up camp. It was early afternoon, so we relaxed, rinsed ourselves and our dirty clothing, and took a nap (a true luxury in the wilderness). That evening, we enjoyed hors d'oeuvres, some Scotch, and a tasty meal of pasta, all while watching the shadows lengthen. Dessert was some chocolate treat, or perhaps a freeze-dried fruit cobbler, I no longer recall—and a sunset light show that I have not forgotten. The meadow grasses turned from auburn to golden, just as shadows overtook the meadow, and then to darkened tan. Chaos and Overlook Mountains provided a complementary show of darker earth tones, from taupe at midday to brown with hints of red in late afternoon, and finally purples as day turned to night. Gradually, colors slipped away, holding on to the fair-weather clouds above for a moment or two, but soon disappearing. The light show had ended, so we turned in, humbled by the grandeur of a meadow virtually unknown.

Overlook Mountain shelters Fishhawk Glacier—or whatever is left of the ice body—on its north side. It's tragically fitting, perhaps, that the glacier is also so close to a mountain whose very name, Chaos, could describe the future climate of the Yellowstone ecosystem—at least compared to the somewhat stable, slow-to-change climate it's known for the last ten or twelve thousand years. Using the Northwest Climate Toolbox again, with Old Faithful as the location (since it's well known and, at 7,300 feet in elevation, representative of the Yellowstone Plateau), we see that Yellowstone's projected climate changes will parallel those in Glacier National Park. Specifically, the annual average temperature will increase an additional 9 degrees by the end of this century, for a total heating of 11 degrees, assuming business continues as usual. This warming will not be evenly distributed throughout the year, but instead will be concentrated in summer, which will see the typical daily high temperature move from a cool 68 degrees (in the 1960s) to a warm 81 degrees by 2100 (see the next few pages for a discussion of the extreme temperatures this summertime warming will bring). Winters will see the smallest change, an annual average increase of 9 degrees, though the typical daily high temperature will move above freezing, to 35 degrees (and winter overnight low temperatures are already higher than at any time in the last fifteen thousand years).[47] Snow will be plentiful but short lived, with winter precipitation projected to increase modestly (from 9 to 11 inches of snow water equivalent, or SWE), while spring overnight low temperatures will move much closer to the melting point (from an average of 19 degrees historically to 30 degrees by 2100). Spring precipitation is also projected to increase, but at these temperatures it's likely to be rain, which will expedite the melt-off. By 2075, most of the Yellowstone ecosystem will be snow free by April 1, the historic peak of the annual snowpack. Finally, summer precipitation, already the lowest volume of the four seasons (barely half that of winter and spring), will remain unchanged, or even shrink slightly.[48] Changes of this magnitude will force several reckonings, in both winter and summer. Let's take a look at the winter first, before proceeding to the summer reckoning.

Winter in Yellowstone is a magical time of year, with snow and cold dominating the landscape. Snow covers everything but the geysers and other thermal features, while also forcing most animals down to the sheltered river valleys. Temperatures well below freezing clash with the hydrothermal warmth, coating nearby trees in layers of frost and rime, keeping most rivers open, and providing places of warm ground that some animals take advantage of on subzero nights. Visitors have long been drawn to such wonders, using snowmobiles and snowcoaches (enclosed, multipassenger vehicles capable of traveling on snow-covered, unplowed roads) to get to Old Faithful and other park destinations. Despite two decades of controversy, both forms of oversnow transportation remain available to park visitors, with noise, emissions, and driver training requirements (the outcomes of the controversy). Most of these vehicles can only be driven on snowpacked roads, a compromise that will prove to be their downfall as this century unfolds. A pair of researchers used three climate models to project the snowfall necessary for oversnow vehicle use. They found that the lowest road open for such use, the West Entrance Road, would only have sufficient snow 59 percent of the time by midcentury, and just 38 percent by late century. The roads branching off from West Entrance Road would also become marginal in some years, declining to 79 percent usable by midcentury and 70 percent by late century. The two authors concluded by urging park managers to consider plowing the affected roads when the time comes, something I suggested in my first book, *Yellowstone and the Snowmobile*. I not only recommended it as a climate change response but also as climate change prevention, given that all oversnow vehicles are gas guzzlers.[49] This idea was analyzed but not chosen in one of the planning documents, largely because the historical, romantic connection between contemporary oversnow vehicles and those first used in the park was too powerful to overcome. Regardless, climate change will soon force this reckoning—one that may happen more smoothly due to the planning that has already occurred in this park.

To get to the evidence that a reckoning will also be necessary in summer, Josh and I needed to return to the fur trappers' thoroughfare, by hiking back down the Mountain Creek drainage. After a frosty start that morning (26 degrees at dawn, after the coldest night of the trip), we bushwhacked along the creek for a few miles before we reentered Yellowstone. Along the way, we came across fresh bear scat, complemented a few miles later by both grizzly and wolf tracks on the trail we rejoined. Signs and sightings of herbivores were also abundant on that hike with Josh, from the two buck mule deer near our camp that night to the two bull bison we saw the next day, as well as the elk bugling we continued to hear at times. I was reminded of a fall hike I'd taken through this area in 1998: camped overlooking the Yellowstone River Delta on the second evening, my hiking partner and I listened to a near-constant serenade of bugling elk and bellowing moose—the ardor of both animals plain to all who could hear—that continued well into the night. It was the loudest encounter I've had with the Thorofare's abundance, an exuberance of wild animals, and wildness, that is rare in America today. Drawing this wildlife to the area is a rich mosaic of habitats, perhaps most obviously on display in the Yellowstone River Valley (the trappers' thoroughfare). Forests of all ages and meadows of all sizes cover the broad valley of the upper Yellowstone, the forests in various stages of postfire maturity (from seedling to senescent trees), the meadows in various conditions of moistness (wet to dry). These diverse habitats make the Thorofare a natural cornucopia of both animal and plant life.

Josh and I took another side trip the next day, hiking up the trappers' thoroughfare. Josh had never gone that way, and it was always a scenic diversion, so we hiked some five or six miles up the valley, had lunch, and then retraced our steps back to camp. Shortly after we embarked on the day's journey, we passed through an area that had been recently burned by the 2002 Phlox Fire. Started by lightning in the Phlox Creek drainage on the side of the Two Ocean Plateau across the valley from us, the fire was initially confined to a patch of trees left unburned by the 1988 fires. However, on the night of August 16, the fire blew up as a dry cold front moved through the park. Driven by sixty-mile-per-hour winds, the fire moved east, into the sapling forest germinated from the epochal fire fourteen years earlier. Young forests like that usually don't burn readily, because they don't have much fine fuel—that is, the sticks and needles necessary for active fires. Driven by that gale-force wind, though, the Phlox Fire moved through the sapling forest, jumping the river and burning across the valley before the wind subsided. It smoldered for another couple of weeks, never escaping the blowout it had burned, which ended at the trail we walked.[50]

For the better part of a decade, the Phlox Fire was a curiosity, a rare example of a young forest that reburned, with the wind understood to be the catalyst for the unusual fire behavior. But in 2011, a group of researchers who had long investigated fire in Yellowstone published a landmark paper arguing that such "reburns" would become common in Yellowstone as the twenty-first century unfolded—and that no wind would be necessary for the reburns to occur, for the warming climate would be the catalyst, with cascading impacts on the area's vegetation. Using three climate models scaled to the Yellowstone ecosystem, the authors modeled its future climate and associated vegetation and fire regime, given current trends. They concluded that years like 1988 will become increasingly common as the decades pass, with one to five of them by 2050 (depending on the model). By century's end, years like that will be the norm, not the exception—and even years like 1988 will seem cool, as days over 90 degrees balloon from none per summer now (even in 1988) to ten to forty by midcentury and forty to seventy by century's end. With large fires dependent on high temperatures and the associated drying of fuels, it's easy to see that the Yellowstone ecosystem is facing a dire future, especially when we factor in the earlier and warmer springs. After 2050, most years will see more than 250,000 acres burn, a result common to all three models the researchers used. In fact, they ran the models a thousand times, and just one of them projected any years without large fires after 2050, and for only three years. Such frequent large fires will be a dramatic reversal of the historical record, where years with large fires were uncommon. The fire interval for a given forest in the ecosystem will drop from one hundred to three hundred years historically to sixty years in 2034, to twenty years or less by midcentury, and finally to ten years or less by century's end. By midcentury, fires will burn so often that they will convert from being climate limited (burning only when years like 1988 come along, no matter how much fuel is present) to being fuel limited (burning only when sufficient fuel accumulates, which is usually five to twenty years, with suitable weather conditions occurring most years).[51] These projections were closely replicated by a researcher who built a climate model specific to Yellowstone, using weather data from stations in and around the park.[52] When similar results are obtained using different methods, researchers consider them highly trustworthy. Consequently, it's a safe bet that if we continue consuming fossil fuels at the present rate, Yellowstone's future will be quite fiery—indeed, the word barely begins to describe it.

As one might expect, such frequent fires are far outside the fire regimes evident in Yellowstone's recent past. When fire moves through a forest, bits of the resulting charcoal are washed into nearby lakes, where some settle to the lake bottom. If the lake has annual influxes of sediment that also accumulate on its bottom (i.e., from seasonal snowmelt), a researcher can extract a sediment core from the lake bottom and methodically go back in time to see when fires burned in the lake's headwaters. A team of researchers has done precisely that with a core extracted from Cygnet Lake in the middle of the Yellowstone Plateau. The core they used was twenty-two feet long and covered the last seventeen thousand years, all the time since the glaciers melted off. Their analysis revealed that at no time in that period did fires burn more frequently than once per century. The longest fire intervals—two hundred to five hundred years—have been in the last two thousand years. Also, lakes preserve the pollen that falls into them, and since pollen can be identified by the plant that produces it, researchers can also determine the relative abundance of plants over time in the lake basin. In the Cygnet Lake area, the only tree to grow during the entire period of record was lodgepole pine, because it is the only species that can cope with the poor soils there, which occur throughout the Yellowstone Plateau. To sum up, since glaciers receded seventeen thousand years ago, the Yellowstone Plateau has not seen anything close to the frequencies of fire projected for its future.[53]

I think it's human nature to put unpleasant things like the landmark paper's projections in the file of things we'll deal with when the time comes, especially if they are so far outside of our experience as to strain belief, no matter the authors' pedigrees or the peer-reviewed journal's reputation. The dates in the article also seemed far in the future, so for five years, the avoidance approach worked well enough. But in 2016, two things happened that brought the paper back into the limelight, forcing the fiery projections back into our attention. The first was the publication of a book that repeated many of the paper's projections, going into more detail about the changes in vegetation the Yellowstone ecosystem would soon experience. The book was expensive and wonky, so it probably didn't do more than reaffirm the projections we all wanted to ignore.[54] The other event of 2016, though, made many observers of fire in the Yellowstone ecosystem sit up and take notice: the Maple Fire. It was to become a portent, at least to those familiar with fire in Yellowstone, that we are pushing nature far outside its ability to cope with our emissions—and that we may not like the changes to come.

On August 8, 2016, lightning started a fire northeast of West

Yellowstone, Montana, in an area of Yellowstone National Park that few people visit. Just north of the Madison River and east of US Route 191, the land is utterly flat and forested with lodgepole pine. One seldom-used trail passes through the area; the few who hike it are rewarded with meadows full of flowers, views of the Madison and Gallatin mountain ranges, natural quiet, and plenty of solitude. The new blaze smoldered for a week or so, but on August 17 exploded into action. Temperatures had been pushing 90 degrees since the first of the month, with relative humidity readings below 20 percent and the moisture content of downed logs below 10 percent—so it didn't take much to spur the fire into action (probably wind). The heat and aridity continued through the end of the month, with the humidity dropping below 10 percent on eight different days. Stoked by the near-perfect burn conditions, the fire marched through the tinderbox forest, sending massive smoke plumes into the dry sky and creating its own weather as cooler air rushed in to replace the air that had gone skyward. The fire burned actively until the first fall cold front brought cooler, moister conditions in early September, though it continued burning well into October, when snow finally put it out. By then, the fire had burned 45,425 acres, the largest area since 1988.[55] Impressive enough, but that's not what the fire will be remembered for. Rather, the Maple Fire will long be known for the fact that it burned almost entirely within the boundary of the 1988 fires—and, for at least four reasons, it may become known as the reburn that inaugurated the wholesale transformation of Yellowstone's forests into drier types, or even nonforest, due to climate change.

The Maple Fire significantly impacted Yellowstone in four key ways. First, it introduced a new fire intensity to the Yellowstone area, burning so hot in places that everything combustible burned, including any serotinous cones that may have been on the twenty-four-year-old trees, along with any logs that had been on the forest floor (most of which were snags and logs left by the 1988 fires). Such fires, which burned for more than an hour on logs, left naked land, nothing but bare mineral soil. These hot spots, termed "crown fire plus" by the research team that reported them, burned in places of forest characterized by very (sometimes incredibly) dense, spindly saplings, less than 1.5 inches in basal diameter. Such thickets were marvels of serotiny's success, areas of such prolific seedling establishment that more than fifteen thousand saplings crowded an acre (for comparison, my parents' one-acre, mostly wooded yard in Missouri has about a hundred largely mature trees). This was the place I took park visitors to see in the 1990s; we marveled at the dense seedlings but knew that such extreme density could never produce a typical forest without some thinning. I remember telling the groups that natural selection would thin them through competition for resources, never dreaming that a second round of fire could play a role in the site's story. Indeed it did, through crown fire plus, though perhaps more radically than I would have thought, since the reburn incinerated the entire seed bank. Reforesting the site is dependent on nearby unburned trees—if there are any within dispersal distance. In this way, crown fire plus may delay reforestation of reburned places, or even preclude it altogether. Similarly, models show that these patches of forest will never regain the level of downed wood present before the Maple Fire burned through, which means the fire catalyzed a new forest structure in Yellowstone, at least in some areas.[56]

Second, fire begets fire, in more ways than we knew it could in lodgepole pine forests. Even before the Maple Fire, a team of four researchers had found that 63 percent of Yellowstone's sapling forests had canopy fuel loads exceeding the levels at which crown fires become active, and 76 percent had sufficient downed logs to sustain a hot surface fire (one that remains on the ground, but usually occurs with crown fires). Once they had 45,425 acres of confirmation that such forests could indeed burn, they modeled the weather and fuel conditions that would be necessary for such fires. They found that not only would the sapling forests burn under the extreme conditions present in 2016 in the Maple Fire area, but they could also burn under moderate fuel moisture and wind conditions. This was a complete surprise to most fire observers, one with potentially dangerous implications for future Yellowstone forests. Today's sapling forests will not be the only young ones to join the Maple Fire in burning, for if fires do increase throughout this century as predicted, an ever-growing percentage of forests in the Yellowstone ecosystem will consist of similar young forests. With more and more heat projected for the area as the century progresses, and increasing amounts of readily flammable forests, it's clear that these two conditions may collide to produce a powder keg just waiting for ignition.[57]

Third, seedling establishment is significantly reduced after reburns, compared with that following the first, stand-replacing fire. Relative to the number of saplings present before the Maple Fire, the number of seedlings after it dropped sixfold. In dog-hair stands of saplings, the drop was much steeper, from around fifteen thousand saplings to fewer than four hundred seedlings per acre. Two-thirds of the aspens that germinated after the 1988 fires were also killed, though those that survived responded with vigorous growth.[58] This low number of both aspen and lodgepole seedlings will probably give rise to a woodland forest (an open forest, with gaps between the trees). Evidence from other reburns in Yellowstone shows that over the span of ten or twenty years after a reburn, other seedlings gradually fill in the woodland gaps. In one hundred to two hundred years, the previous woodland might converge in function and appearance with a postfire forest that never saw a reburn, such that even a trained forester would be unable to discern a difference.[59] Nonetheless, the fact remains that seedling density is low following a reburn, which makes the ensuing forest vulnerable to disruption by destructive events—including, or especially, those caused by climate change.

And fourth, there is abundant evidence that the warmer and drier conditions caused by climate change are already compromising seedling survival following fires, and these effects will only get worse as climate change intensifies in the decades to come. The year 2000 marked a significant change in seedling survival throughout the West, caused by a 14 percent increase in the annual water year deficit beginning that year. This change was paralleled by an increase in recently burned sites in which regeneration failed (from 19 percent to 32 percent), along with a drop in the percentage of sites in which the number of seedlings established after a fire equaled or exceeded the number of trees before the fire (from 70 percent to 46 percent). All forest types were affected, especially the driest ones dominated by ponderosa pine and Douglas fir.[60] In another paper, a team of researchers surveyed burned areas throughout the West, finding that Engelmann spruce and subalpine fir establishment after fire declined sharply when drought struck after the fire.[61] Even the survivor of survivors, lodgepole pine, is expected to struggle as the century progresses. Seedlings subjected to the climate conditions they will likely encounter midcentury experienced a 92 percent reduction in establishment following fire (Douglas fir seedlings, included in the same study, had a 76 percent reduction).[62] In a modeling exercise, the same two trees failed to regenerate thirty years after a hypothetical stand-replacing fire, with Douglas fir regeneration failing 55 percent of the time, and lodgepole pine failing 28 percent in non-serotinous stands but just 16 percent in serotinous stands.[63] Finally, lodgepoles at their lower, present-day treeline will, by midcentury, find the climate too warm and/or dry to successfully reestablish after a stand-replacing fire.[64] With whitebark pine providing a preview for this disruption, it's clear that most Rocky Mountain trees will find regeneration difficult or impossible after future fires, especially in the case of reburns, due to climate change.

For these reasons, the Maple Fire may well provide us with a preview of Yellowstone's fiery future (especially because it burned at the lower treeline for lodgepole pine in the Yellowstone ecosystem). If the crown fire plus sites don't reforest, if the few seedlings present there today perish in a drier and hotter summer, and especially if another fire burns through before a robust seed bank is established (i.e., before 2046), some of the 45,425 acres could well pass a tipping point, failing to regenerate and becoming something different entirely. Indeed, it is hard to overstate the calamitous effects that these radically enhanced fires will have on the forests of the Yellowstone ecosystem. Fires will be the catalyst of the changes wrought by the warming climate, forcing the tree species to move upslope where possible, following the climate they prefer up the mountainside. In their place will move species that can handle frequent fire, such as ponderosa pine. This tree, currently restricted from most of the ecosystem by summertime frost,[65] produces thick, corky bark that insulates its cambium from the heat of low-intensity fires (like Douglas fir, but ponderosa pine can tolerate drier conditions). Other forests will transition to a mix of sagebrush-juniper vegetation types, which will increase in coverage from 41 percent to 71 percent of the ecosystem. Included in this grouping are grasslands and open woodlands with scattered ponderosa pine, Douglas fir, western larch, juniper, and even Gambel oak (a scrub oak new to most of the ecosystem), along with Great Basin desert scrub, the sagebrush-dominated plant groups that cover Nevada and western Utah. Almost unknown in the Yellowstone ecosystem now, desert scrub will find almost half the ecosystem (48 percent) suitable for growth by century's end. Conifer forests will fall from 41 percent of the Yellowstone ecosystem to just 19 percent by 2100. The main losers will be the subalpine trees: Engelmann spruce, subalpine fir, and especially whitebark pine. The three will likely persist as young forests on the highest mountains, and in refugia like north-facing slopes where the local climate is cooler and moister. The relative winners will be the open woodland trees, but the resulting landscape cover will have little resemblance to the current one.[66] Vestiges of today's forests will persist in areas passed over by fire (in some cases, for decades), but an increasing portion of the landscape will consist of vegetation types new to those locations.[67] By century's end, the forests of Yellowstone will largely be gone, the vegetation virtually unrecognizable to those of us who know it now.

Back in Missouri, I began going to an ALS clinic every other month. As with the clinic I had been attending in California, a variety of the medical specialists an ALS patient needs all assemble in one location for an afternoon clinic, rotating through the rooms in which the patients sit. My parents and I quickly grew weary of the Friday afternoons we spent there. Not only were those afternoons tiring (the typical visit was three hours long), but there was never good news. Always, the disease continued its inexorable march of destruction through my body's musculature and nervous system. The most telling measure was the breathing tests I took every time we went there, which consisted of blowing air as hard as possible into a device that measured the volume and pressure of my exhale, and doing the reverse for the inhale. The test results were given as a percent of that expected for a person of my build, putting hard numbers on my decline. The percentage was 91 when I first took the test shortly after the diagnosis and was still in the 80s when I arrived in St. Louis, but continued its steady decline there, about 1 percentage point per month. A year and a half later, it was still in the 60s, but between two clinic visits, the rate of decline jumped exponentially, to 10 percentage points per month, dropping me into the mid-40s. It returned to the reduced rate of decline the next month—a rare example of good news—but even that was tempered by the fact that I did not recover the lost ground. For this reason and many more, we often shed tears in the evenings after the clinics.

The breathing test results not only provide a measure of an ALS patient's decline, but they also serve as a signal to the neurologist as to when a patient needs breathing assistance. When the results drop below the 50 percent mark, the doctor usually recommends the patient consider using a bilevel positive airway pressure, or BiPAP machine, at night. The BiPAP machine is a device that forces more air into a person's lungs when they inhale (it's the same device as the CPAP—continuous positive airway pressure—machine that people with sleep apnea use, just on a different setting). I reluctantly agreed to try it out, and although it took a full month to get used to having the face mask on me, I've been a regular user of it since then. The decision about whether to use it is the first of many that ALS sufferers must make. Most do decide to use it, in part because it's not invasive and it does help. A minority of sufferers see their breathing ability decline faster than the rest of their muscles, and are confronted with end-of-life decisions much earlier than other sufferers. Some choose to have a tracheostomy performed, which installs a breathing tube into the trachea through a hole made in their neck, and then live out their remaining time connected to a breathing machine that does all the work for them. It's an invasive solution that makes speech impossible, so many, if not most, choose to limit their lives to what a BiPAP machine (or similar device) permits, which is what I have chosen. Most sufferers don't have to make this decision so early in the progression, and my choice might have been different had I been one of those doubly unfortunate persons. The disease, though, leaves no one absolved from ignoring the decision forever—ALS is 100 percent fatal, and about 80 percent of its victims die of respiratory arrest.

In the Thorofare, after our day hike up the Yellowstone River Valley, Josh and I arose to find frost on the tent again, the dawn temperature right at freezing. We crossed Mountain Creek to cook breakfast in the bright sunshine on the far side, our campsite stubbornly cold in shadows the sun wouldn't reach for several more hours. Our fast broken, it was time to retrace our steps out of the Thorofare and to our canoe, nine and a half miles away. Fording Beaverdam Creek, we saw some frogs that had just metamorphosed from tadpoles, and then looking back into the Thorofare from Terrace Point, we had almost a bird's-eye view of acres upon acres of good amphibian habitat in the ponds and wetlands of the delta. That any amphibian survives in Yellowstone's cold climate amazes me, but two frog species, two toad species, and one salamander species live in the park. Most likely, we saw Columbia spotted or boreal chorus frogs, both of which are widespread, with many breeding locations in the Yellowstone ecosystem. Perhaps their most important adaptations to this cold environment are those that get each species through the long winter; Yellowstone's amphibians display at least three ways to cope with six months of snow and cold. The Columbia spotted frog finds water that doesn't freeze, and toads and salamanders enter underground burrows, but the chorus frog does something truly marvelous: it freezes solid, stopping its heart and all other bodily functions, in sheltered spots like the insides of fallen logs. Whatever the overwintering strategy, these amphibians are well adapted to an unforgiving climate.[68]

As expected, the advent of extreme global warming will put all amphibian adaptations to the test, along with their ability to evolve, because these very habitats (wetlands) are most at risk in a warming climate. Already, up to 40 percent of shallow wetlands in Yellowstone and Grand Teton National Parks disappear in dry years like 2007, and many wetlands in Yellowstone's lower elevations have disappeared entirely. The increasing frequency of dry years and the much warmer summer temperatures projected for later this century will only increase evaporation, exacerbating the loss of such wetlands. Deeper pools will persist longer, for many of these wetlands are connected with groundwater or other more persistent water sources. Drying of shallow wetlands will impact chorus frogs the most, since they prefer ephemeral pools and shallow ponds. However, spotted frogs, which prefer deeper water bodies, may also be at risk, for sustained periods of low runoff have already led to 40 percent reductions in their occupancy elsewhere in the West. Really, since all amphibians depend on bodies of water for breeding and tadpole development, they are all vulnerable to climate upheaval. So are many other species that depend on wetlands for parts of their lives, including moose, beaver, trumpeter swans, and sandhill cranes.[69]

Native fish are also vulnerable, due to both the projected declines in streamflows and the warmer water temperatures that come with a hotter climate. The sixty years from 1950 to 2010 saw streamflows reduced in 89 percent of central Rockies streams, which includes those in the Greater Yellowstone Ecosystem. Consistent with climate projections, streamflows were most reduced in summertime, especially in the Yellowstone River. Water temperatures in Yellowstone's streams rose by 1.8 degrees in the past century, with the rate of warming in the 2000s greater than that seen during the Dust Bowl. If we continue business as usual, stream temperatures throughout the Yellowstone ecosystem will climb an additional 1.4 to 3.2 degrees by midcentury. Such warmer water would lead to a 26 percent drop in the Yellowstone cutthroat trout population, due to competition from more warmth-tolerant non-native species.[70] Yellowstone cutthroat trout have already been hit hard by whirling disease and by lake trout illegally introduced to Yellowstone Lake in the 1990s. Despite an aggressive gillnetting program to control lake trout there, the cutthroat trout population remains suppressed, a small fraction of what it once was. Climate change will only make things worse for a species once so abundant that they numbered in the millions in Yellowstone Lake alone.[71]

Animals that depend on abundant snow, such as wolverines and Canada lynx, are also vulnerable to climate warming. While both of these species are very rare, they are present in the Greater Yellowstone Ecosystem. The lynx are exceptionally rare, with only 112 reliable sightings in Yellowstone's long history. These felines are well adapted to deep snow, with large feet that serve as snowshoes, long and dense fur, and the ability to roam far and wide in search of food (and a mate). In winter, lynx subsist almost entirely on snowshoe hares, lagomorphs that prefer old, unburned forests. This habitat need alone will be challenged in Yellowstone's fiery future, but climate change poses another significant hurdle to the hare—a growing mismatch between the animal's camouflage and the background. The rabbit molts twice a year, in spring from white to brown, and in fall from brown back to white, to match the typical background of the ensuing summer or winter season. Both molts are triggered by changes in the length of daylight, which closely matches the landscape's conversion from snow covered to open land and back again. If climate change reduces the length of winter at both ends (fall and spring), the hare will lose its camouflage, being white against a brown landscape in both spring and fall, molting too early in fall and too late in spring. While this may seem like easy pickings for lynx (and therefore a boon), it would also make hares more vulnerable to predation by other carnivores, like coyotes, wolverines, and foxes. Whether snowshoe hares could withstand the additive predation is an open question, as is whether lynx could contend with the likely reduction in their preferred prey.[72]

In contrast to the cats, wolverines are more opportunistic, scavenging where possible and hunting just about anything that moves, even animals much larger than itself, like deer, elk, and even moose (the animal's reputation as a fearsome predator is well deserved). Wolverines also den in deep snow to give birth and nurse their young. Monitoring projects in the 2000s detected seven wolverines in and around Yellowstone; two of them were females with home ranges of about 175 square miles, while the remainder were males, three with home ranges twice the size of the females', and two that dispersed out of the ecosystem (one of which became the first wolverine in Colorado in ninety years). Two of the wolverines were in the Thorofare.

Both of these snow-dependent species are either listed as threatened under the Endangered Species Act (Canada lynx) or are candidates for such listing (wolverines), pending more information about their status. The main threat to both animals is climate change, which will limit the all-important snow, but wolverines will also find the projected increased summer high temperatures too hot, much like the pika (and possibly the lynx, though I was unable to confirm this).[73]

These two predators, along with the amphibians, can be considered specialists in that they need specific habitats for their homes, whether those are water bodies or snowy subalpine forests. Specialists are inherently vulnerable to ecological change, limited as they are to specific places that provide the shelter, food, and water they need to survive. In contrast, generalists find their needs met in a variety of habitats. Elk, for example, eat a variety of different plants, from grass to forbs (wildflowers like glacier lilies, dandelions, and spring beauties), shrubs (mostly willows), and tree bark (mainly aspen). They find these plants throughout the Yellowstone ecosystem, along with forests that give them shelter and abundant water sources. Generalists tend to cope with environmental change better because their needs are more plastic. For this and other reasons, generalists are expected to do better as the effects of climate change become more intense. For example, grizzly bears, black bears, wolves, cougars, elk, mule deer, and bison will all probably remain in the Yellowstone landscape for decades to come, though their populations will likely fluctuate and their range may shift. Specialists, meanwhile, will vary in response to the upcoming changes, with those that prefer habitats that are likely to expand holding steady or increasing in number, and those dependent on habitats that are projected to shrink falling in number or dying out altogether. Pronghorn, for example, will likely increase within Yellowstone as the sagebrush steppe habitat they prefer expands, while moose will probably drop in number or go extinct in the region as the wetlands and old-growth forest they need shrinks in coverage. These are broad generalizations based on my training in ecology and my intimate knowledge of the Yellowstone area and its wildlife, so there will undoubtedly be exceptions to them—along with surprises, good and bad.

Moose provide an unfortunate illustration of how animal populations may succumb to our warming planet. Once so abundant in the Thorofare that one observer tallied seventy-six sightings of them in one day, moose numbers have fallen from over a thousand in Yellowstone in the 1970s to fewer than two hundred today.[74] The 1988 fires burned much of their winter forage (the needles of subalpine fir and Engelmann spruce), hitting the population hard, but their numbers have continued to fall since then. While biologists have found a variety of proximate causes, many are suspecting climate change as the ultimate culprit. One of the immediate causes is an explosion in winter tick numbers that, in turn, is facilitated by warmer winters. Individual moose have been found with fifty thousand to seventy thousand ticks on them, about ten to twenty times the normal number. The ticks irritate the animal's skin, so they will rub them off, but that also rubs their fur off right when they need it most, making them vulnerable to exposure. Another factor in the decline is heat stress, especially on warm winter and spring days, when moose (still) have a winter coat on that keeps them warm at -20 degrees. To avoid overheating, the animal will bed down at times when they would ordinarily be feeding. Finally, fires and predation take an ongoing toll (fires by burning forage, not the moose itself, which rarely happens), and with fires projected to increase as much as they are, the outlook for moose in the Yellowstone ecosystem is not good. All of these factors in the moose decline can be linked to climate change, so it is almost certainly the underlying cause.[75]

For six of my twenty-two years in Yellowstone, I gave full-day tours of the park, in both buses and cars. Few jobs could have been better: I got to immerse myself and the visitors in the wonders of Wonderland. We'd ply the park's roadways, connecting the dots between Old Faithful Geyser, the Lower Falls, Yellowstone Lake, and the park's many other attractions. Along the way, we'd run into traffic jams, the telltale sign that a representative of Yellowstone's charismatic megafauna was nearby. Based on the habitats in the area and my knowledge of the park, I could usually guess what quadruped was responsible for the traffic jam (which would then lend its name to the holdup, as in a "bison jam" or an "elk jam"). Also, based on the behavior of the park visitors outside our vehicle, I could usually tell if it was one of the three most coveted wildlife sightings. If visitors were leaving their car doors open, it was probably a grizzly bear. If they were also leaving the engine running, it was probably a wolf or a pack of them. And if they were also leaving the kids behind *and* acting like they were high on mushrooms, it was probably a moose. While I write this with tongue in cheek, it was always clear that visitors adored moose, much more than the other ungulates we associate with Yellowstone—even bison, which many repeat visitors consider the park's most iconic ungulate. Moose had it made as one of the three most desired wildlife sightings, the only one that wasn't a carnivore.

Despite the awe they inspire, we're well on our way to wiping moose out in the Greater Yellowstone Ecosystem, still another example of the self-inflicted harm we're causing. For an unknowable number of species in addition to moose, climate change will cause a series of ecological changes that will make it increasingly difficult for them to survive in the Yellowstone ecosystem—it will result in death by a thousand cuts, for the entire species. The disappearance of moose will be noticed (their increasing scarcity is already being widely noticed), but who will notice when the wolverine and the lynx and the pika disappear from that incredible landscape? Or when the chorus frogs no longer serenade us in spring? Very few of us will take note, of course, but we will all be the poorer as the Yellowstone ecosystem's web of life begins to fray. Every species lost is one more thread of life and ecological connection destroyed, one more part of an ecological whole gone, perhaps forever.

And there are so many signs that climate change is already undoing Yellowstone's rich tapestry of life, signs that began at least three decades ago with the 1988 fires. They catalyzed an upward shift in the lower treeline as the trees there found conditions too warm and dry. In the years since, alpine tundra has warmed so much that it's no longer tundra, moose numbers have fallen dramatically, and whitebark pine has succumbed to bark beetles by the millions across the ecosystem. Driving these changes is global warming, with warmer winters playing a key role in many of the changes we're seeing. Moreover, less snow is falling and snowpacks are melting out earlier, shortening winter by a month and lengthening fire season by twice that (and then some). Forests are responding as we saw in 1988, though nature had a significant role in those eventful blazes. However, the Maple Fire suggests that nature's role may be declining in relation to our own, in determining whether a forest burns, how severely it burns, how quickly it burns again, and what the regenerative forest looks like—if it reforests at all. Indeed, if the Maple Fire is an indication of the future Yellowstone, it will display the hands of humanity as much as it did the hands of nature when I spent my first summer there, in 1986. What's more, if the research base in the other parks examined in this book were as robust as Yellowstone's, we would probably find a similar—and growing—human influence on their forests (especially in

Glacier, with its similar forests). Truly, we are taking the national treasures into uncharted terrain.

The loss is more than simply biological, especially with moose, which are so much a part of visitors' hopes and imaginings. It's about the most fundamental element of the Yellowstone landscape, the wildness found in precious few places in America today outside of Alaska. It's that intangible something, the feelings one senses in a landscape inhabited by wild animals, especially those larger than us. It's the intrigue of encountering something unexplainable, like the ability of trumpeter swans to find their way back to the same obscure pond, year after year. And it's the feeling one gets deep in the wilderness when a grizzly bear looks your way, but seemingly looks right through you, as though you don't exist—and definitely as though you are not the resident or homeowner there. Others call this attribute mystery or wonder; whatever the label, Yellowstone exudes it now. But with every animal gone, every degree added, every snowfall turned to rain, this vital essence diminishes, making it more and more difficult to sense, turning Yellowstone increasingly into just another western landscape, influenced as much by humanity as by nature. It's a matter of degrees, in more ways than one.

Josh and I finished the day with a forty-five-minute paddle to our last campsite, the same one we had used the first night of the trip. That evening, the smoke we'd noticed in the air intensified, producing a red sunset that, given what I've learned about climate change there since, could easily be regarded as a portent of the region's future. Heat, drought, and fire may well come to define the twenty-first century in Wonderland. The altered sunset could also signify my own turbulent future, for the Eagle Peak trip became my last deep dive into the Thorofare. Never again would I walk the fur trappers' thoroughfare; indeed, the closest I would ever again get was my last wilderness trip, in 2014, a year after being diagnosed (the trip I chronicle in *A Week in Yellowstone's Thorofare*). Had I anticipated either future on the Eagle Peak hike, I would have lingered longer—much longer—at Terrace Point, memorizing the sights, sounds, and feel of that powerful place. In just seven years, I would experience the first symptoms; in just eight, I would receive the diagnosis of the most feared neurological disorder; and in just ten, my legs would crumple under me, no longer able to support my own weight, inaugurating the next phase of the disease—life in a wheelchair.

Just before I was forced into a wheelchair, I met a woman who had chosen to have a tracheostomy done. Anne Rayburn was one of those people who saw her breathing ability drop markedly after diagnosis.[76] Just two years into her disease progression, she was hospitalized with breathing difficulties, which led to her decision to get the trach (as they are known in the health-care world, rhyming with "rake") tube put in. In another three years, she saw her motor neurons mercilessly erode, to the point that the only muscles she could use were her eyes and eyelids. Her only way to directly communicate was through the speech device that she ended up lending to me. More often than not, she relied on her husband, George, asking yes-no questions that she could answer by looking up with her eyes if her answer was yes, and down if it was no. Fearing she would lose even those forms of communication, George asked her if she wanted to continue living if she lost control of her eyelids and eyes. Believing a cure for ALS was soon to be discovered, she raised her eyes. Not long thereafter, her eyes could not open, completing the disease progression and leaving her in a most unimaginable existential state. With absolutely no way to communicate, she entered into what can only be described as a hellish form of solitary confinement. ALS does not affect the senses, so she was still capable of feeling pain—hot and cold, headaches, and the innumerable other discomforts with which most of us deal without a second thought. Being imprisoned in her own body, though, she had to suffer through till it went away on its own or something else more painful and intense came along. If she had a headache, there would be no aspirin; if she had a stomach cramp, no antacid; pain on urinating, no antibiotic for the probable urinary tract infection—and on and on. Using my own experience, I sometimes get uncomfortable at night, like we all do, and I need to be repositioned. That is something I cannot remedy myself, so I call for my parents on a room monitor, and one of them soon comes to help. For many reasons, sometimes it takes them a half hour to register that I'm calling, during which time the discomfort can become pain, sometimes sharp pain. For Anne, that pain would only grow, hour after hour, sometimes day after day. Being in her kind of solitary confinement must have been a seemingly endless sentence of horrors.

Meeting her and George at their home in Imperial, Missouri, left me with several indelible impressions. The first occurred just from seeing her sitting prone in her wheelchair, mouth agape, immobile tongue hanging out, all because her jaw muscles were paralyzed too. The second impression happened when we went out to their deck to see the Mississippi River in the distance. While we were outside, George opened her eyes with his fingers. Her eyes were locked in, looking unwaveringly forward, not looking around as any of us would typically be doing. Lastly, for the remainder of the evening, she sat with us, unmoving, unsmiling, unspeaking, unlike any person I could recall encountering. Her time in solitary lasted about two years. At the funeral parlor, her mother told my mother that my decision not to have a tracheostomy done was the right one. Indeed, when her eyes closed for the last time, you could say that she essentially stopped being a human being, for she was unable to do many of the things that make us human, like talking, laughing, and participating in the human community. Anne's situation took that isolation to the extreme, but ALS does this to all its victims. ALS is truly one of the most dehumanizing maladies afflicting human beings, as Anne epitomized.

Finally, I wonder what Anne used to pass the time and distract her attention from whatever daily pain she felt. She was a devout Christian, so it's a safe bet that prayer occupied her thoughts some of the time—and probably more for others than herself. She had kids, so they surely were in her thoughts some, as was George. He says she adored all children, working with them much of her life, so undoubtedly those other kids were in her mind too. Beyond these diversions, what were her other pastimes? Did she have any secluded places of serenity, places where she could feel relief from the constant pain, if only for a moment? Did she envision herself, as I would, on the shore of Yellowstone Lake, watching the setting sun highlight a departing storm while the tempestuous waves at my feet slowly calm? Could she sit on a fallen giant in Sequoia National Park's Log Meadow, feeling the exuberance of summer life in that spot of heaven? Or did she recall the timeless peace and serenity of the river trip she took through the Grand Canyon, transported by the river through multiple dimensions of time, space, and beauty? I hope she had some such place of tranquility and natural beauty, even if it was only the view from her deck. For myself and millions of others, a special area in some national park is that refuge, that virtual port in a personal storm. This is, perhaps, the most important reason to prevent the worst of climate change: to save our places of wonder and inspiration, our citadels of safety, our retreats of renewal and transcendence. In saving our natural cathedrals, we'll save ourselves too.

Chapter Five

YOSEMITE NATIONAL PARK

Yosemite National Park in California preserves some of the grandest mountain scenery of the West, including the incomparable Yosemite Valley, the cloud-scraping Sierra Nevada mountain range, and the rugged canyons of the rivers draining the range. Yosemite Valley—that seven-mile cleft in the mountains that juxtaposes a flat, one-mile-wide series of meadows and forest against cliffs that virtually soar into the heavens—is as close to nature's cathedral as any earthly paradise. Five major waterfalls grace its walls, including the tallest waterfall in North America, Yosemite Falls, whose total drop is 2,425 feet, with Upper Yosemite Fall a free fall of 1,430 feet. The Sierra Nevada, John Muir's "Range of Light," forms much of Yosemite's spectacular high country. Composed mainly of granite, the Sierra gives rise to streams and rivers whose water is so clear, I've often thought of it as liquidized air you can almost inhale.[1] With two vertical miles of topographic relief, Yosemite has habitats ranging from near desert in its lowlands to alpine on its mountaintops, with an exceptional variety of plants to match—1,565 species.[2] Star among the plants is the giant sequoia, the largest tree on earth. Yosemite has three groves of the burly trees, including the Mariposa Grove, first protected by President Abraham Lincoln in 1864 (along with Yosemite Valley). Bigness doesn't stop at giant sequoias; the park also preserves some sugar pines, the largest and tallest pine species in the world (up to 11.5 feet in diameter and 274 feet tall), with the longest cones (up to 2 feet long). Yosemite preserves the largest of these largest pines, along with the largest white fir, red fir, and (nearby) sierra juniper.[3] Tree diversity is big as well; Yosemite has as many pine species alone (nine) as Yellowstone has conifer species total.[4]

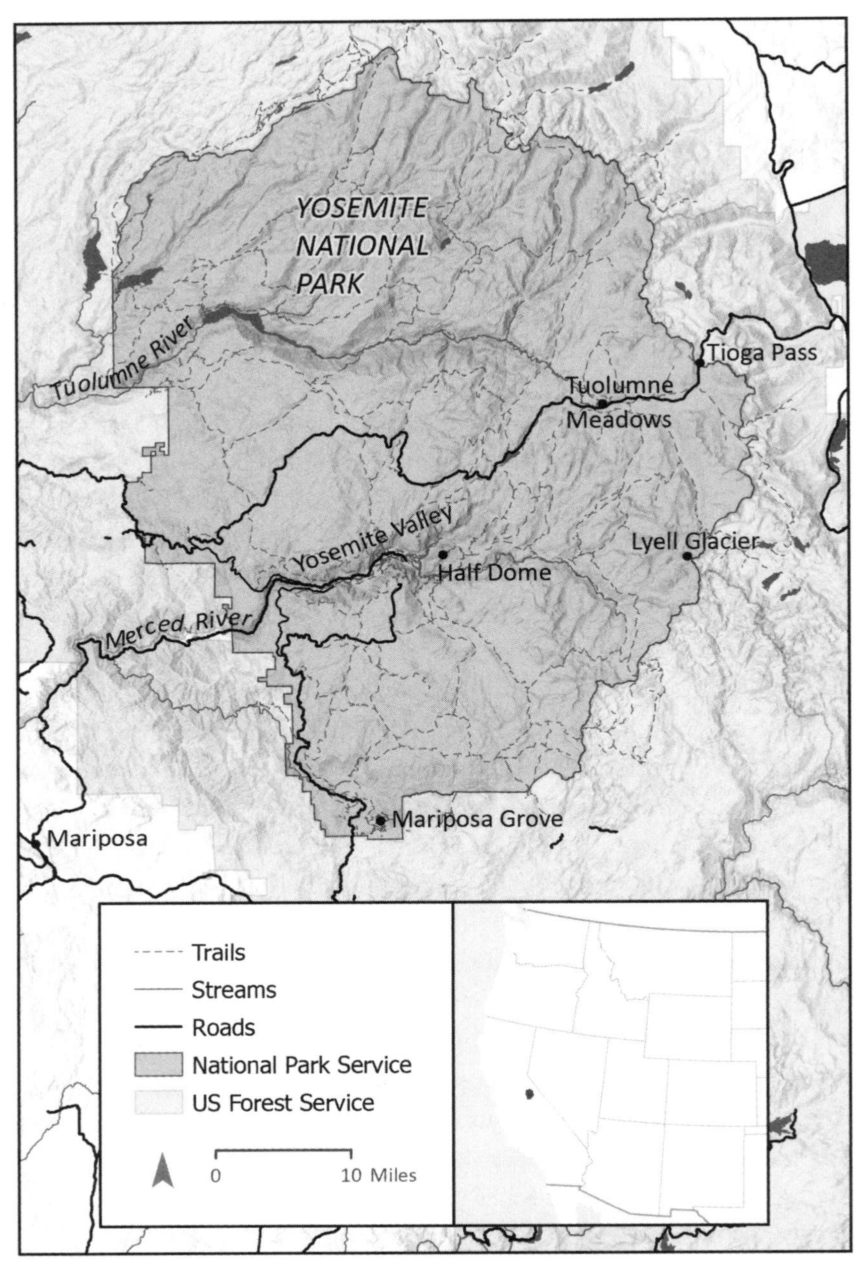

MAP 5 Map of Yosemite National Park. *Sources*: NPS, USGS, Natural Earth.

With its biodiversity matching the grandeur of the incomparable valley, Yosemite is indeed a national and international treasure.

I first became acquainted with the Sierra Nevada's wonders in 1989, the summer after I graduated from college. I had landed an entry-level seasonal position at Sequoia National Park's headquarters, located in the Sierra foothills (Sequoia is just south of Yosemite, but still in the Sierra Nevada). Cool and wet in winter (usually), this low country quickly introduced me to the other half of its Mediterranean climate: hot and dry, for months on end. Daytime highs were always in the nineties or low hundreds, lending expediency to my weekend escapes to the higher ground of Giant Forest and Mineral King, the cooler provinces of the giant sequoias and High Sierra, respectively. After a second summer, in which I was stationed at Giant Forest, I left the Sierra and California, I thought for good. Although I had developed a special fondness for the big trees (particularly one lovely meadow in Giant Forest), California was too crowded and polluted—even in Sequoia—for my taste, so I returned to Yellowstone. Nineteen years later, though, I found myself returning to the Sierra to accept a significant promotion in Yosemite. There, I led the development of a management plan for the Tuolumne River, including an exquisite set of meadows along the river. It wasn't long until those meadows occupied as cherished a spot in my heart as the one in Sequoia, became a place that stirs the soul and imagination, a place that soothes and comforts, a place of constancy and stability.

I last saw Tuolumne Meadows in July 2014, when I flew into California from Montana for the ALS clinic at Stanford University—the one where I heard that I needed a wheelchair. After that ignominious experience, I sought solace with Yosemite friends, who joined me on a two-day camping trip to the meadows. We sat by the Tuolumne River, caught up with each other, sat around the campfire, and had dinner at the historic Tioga Pass Resort outside the park (built in 1916) one night and at the non-historic-but-well-on-its-way-to-being-historic Tioga Gas Mart / Whoa Nellie Deli (in Lee Vining) the other night. We made memories together, experiences that are now as indelible as being told I need a wheelchair. And now that I'm in that very wheelchair, I not only draw strength and solace from those enduring impressions, but I also revisit those places and moments in my dreams and daydreams. My dreams emphasize the dominant impressions—and anxieties—I have from my times in the park, so just as I relive the colors of Glacier, I fret about development in Yellowstone's Thorofare. Similarly, my Sierra Nevada dreams sometimes include crowds, usually in the backcountry, but others feature liquidized air and big trees.

For this chapter, then, let's relive a spring trip I took to Yosemite's Tuolumne Meadows in 2011, a cross-country ski trip that I can only dream of repeating today, for my travel days ended when I moved to St. Louis three years later. The trip took us past evidence that the Sierra climate is already changing, with a dramatic example of those changes beginning just a year later: a four-year drought that shattered many records. This drought was not just one for the record book, but it was also an uncomfortable peek into the future for Yosemite in specific and for California in general, a glimpse of a future that could easily be more of a nightmare than a happy daydream.

One morning in April 2011, five cross-country skiers, one of them myself, met at the airport in Mariposa, California, the first town of substance outside Yosemite's western entrance, not far from the house I rented at the time. We were assembling for a quick flight across the Sierra Nevada to the tiny hamlet of Lee Vining, from which we would ski back to the two cars we'd parked in Yosemite Valley the night before. The flight would take a little more than an hour; the skiing back over would take four days, plus a layover day in Tuolumne Meadows. (Driving around the Sierra to Lee Vining in winter would have taken us seven hours, thus the flight.) We were three guys, plus a sixty-two-year-old woman and the guide, another guy, by the name of Terry. I don't remember the names of the others, so we'll call the woman Jean, the two other guys Bruce and Dave. The trip was sponsored by the Yosemite Winter Club, a group of skiers, both downhill and cross-country, that puts together trips like this and other social events. We would stay in park cabins spaced about a day's travel apart from each other, where we would find food and drinks stashed by club members months before—copious amounts of both, since the club ordinarily saw more people sign up for the Trans-Sierra trips than for any others. We'd had to postpone this trip due to a winter storm, which meant some people had dropped out, unable to change their schedules. Doing the trip in April meant we would complete the ski journey when the snow was its deepest, which would be fun to see. It had been a heavy snow year, with one week in January bringing three major storms through the Sierra, dropping an average of 10 feet of snow above 7,000 feet throughout Yosemite. (Coming from the snowy climate of Yellowstone, I was accustomed to a big winter storm dropping up to a foot of snow, but in the land of bigness, storms are not considered big until their snowfall is measured by the yard!)

With all our gear stowed in the two small planes we had chartered, we settled into our seats and took off. It was a clear day, so the pilots flew low, catering to their passengers' sightseeing desires. The foothills around Mariposa were green with the warmth and moisture of spring, but it wasn't long till we were looking down on Yosemite Valley, rimmed by its white-capped Valhalla cliffs. We peeked up the valley that gives rise to Bridalveil Fall, looked down on Yosemite Falls and its giant snow cone (built up over the winter at the base of the upper fall), and then looked back on Half Dome, the plane whisking us onward to the high country. With the landscape below us getting ever whiter, we looked ahead—and still up, even from our height—to the looming Sierra crest, a solid line of mountain peaks scraping the sky. Fronting the rugged lineup was a nearly unbroken field of tundra, a flatness of glistening white that perfectly offset the peaks above. As we got closer, we saw the tundra drop off to the valleys of the Merced and the Tuolumne—the two rivers draining much of our viewshed—and all of Yosemite. These rivers and valleys retain their grandeur today, but they were even more impressive in the not-too-distant past, as we would see throughout the trip.

Indeed, we soon saw evidence that they were rivers of ice not so long ago, glaciers that sculpted the land below us. Just shy of Tuolumne Meadows, the pilot banked to give us a bird's-eye view of the Cockscomb, a jagged, knife-edge ridge of granite spires breaking through the blanket of glitter. One of several similar ridges and peaks south of the meadows, the

Cockscomb and its neighbors are arêtes, those glacially carved, parallel sharp ridges we first met on Mount Olympus. True to form, the ridges are all aligned in the same direction, northeast to southwest, giving away the glacier's onetime direction of travel. Here, the glaciers began on the Sierra crest, flowed down to Tuolumne Meadows, and backed up against the Cathedral Range (the mountain range encompassing the group of arêtes). Piling up there to a depth of two thousand feet, the glacier turned west, carving the Grand Canyon of the Tuolumne River. More ice accumulated in the arête field, coalescing into another glacier that moved southwest and downward, eventually joining forces with a third glacier to gouge out Yosemite Valley. The Cathedral Range glacier left its calling card in the form of erratics—boulders that don't match the underlying bedrock on which they rest today, in the Yosemite Valley area, but do match the bedrock of the Cathedral Range, from which they'd been plucked. The view from our plane window was thus not much different from what it would have been twenty thousand years ago: a landscape of white, being shaped and molded by snow and ice.[5] We were about to step into a world that will soon lose most or all of that defining influence, as we'll see clearly as the trip goes on.

An hour or so later, we were on the ground and at the gate where the California Department of Transportation (Caltrans) closes the Tioga Road for the winter. We'd flown out of the park and over Mono Lake, landed at the Lee Vining airstrip, and gotten a ride to the gate from some associates of Terry's. The ground was bare, the temperature in the fifties, and we had seven or eight miles to go and three thousand feet to climb, so we didn't linger. Hoisting our packs with the skis strapped to them, we began the trek up to the snow, which was about halfway up the climb. About a third of the way to the snow, a Caltrans employee driving a pickup pulled up beside us and stopped to chat. The agency had begun to clear the road as far as the pass for the summer opening, but its rotary plow had broken down. He was on his way to fix it and offered us a ride, though he had room for just two others in his rig (and would not let us ride in the bed). We gallantly let age and beauty take the offer, so Jean and Dave climbed in while the rest of us threw our packs and skis in the bed of his truck. The Caltrans mechanic came back to get the rest of us, so within another hour we were on snow, with less than half of the climb remaining. We ate our lunch, said thanks to the kind man, and strapped on our skis. It was time to reenter winter.

And what a winter it was! The April 1 snowpack measurements (which provide the most comprehensive assessment of the winter's total snow water equivalent, or SWE) had just been tallied, showing the Merced River drainage to have a snowpack that was 172 percent of average, with the Tuolumne drainage at an even greater 178 percent. For both drainages, that's roughly tied for second in the fifty years from 1970 to 2019 (first place is 1983, a legendary winter that brought well over twice the average snowfall).[6] If the five-foot step onto the snow by the plow was not illustration enough of the winter's largesse, the almost completely buried buildings at the Tioga Pass Resort certainly were. The snow was above the eaves of most cabins there, leaving only the attic vents at either end of the buildings visible under the thick blanket of white. A few smaller buildings were totally buried, just mounds of snow indistinguishable from a buried boulder. Even the main lodge was submerged in SWE, with small windows taking the place of attic vents. Recognizing an opportunity for fun and memorable moments, I broke a trail to the roof crown and posed for a photo. Rather than trying some daring descent that could snap a tendon or ligament, I took a safe route down, thereby demonstrating that I had outgrown most of the testosterone poisoning that causes young men to do the stupid things in the *Death in [name your National Park]* books that are becoming increasingly popular. Not to be outdone in the manly contest, the other members of our group followed my tracks to the rooftop and back down—everyone, that is, except Jean, who must have been thinking that, once infected, the inferior sex never truly recovers from testosterone poisoning.

Be that as it may, we still had two miles to go, with only two hours of daylight left, and Jean was skiing slowly. At about six thirty, the faster skiers and I arrived at the pass, with the others arriving within an hour. Right at the pass was the entrance station, a historic building designed by NPS architect Herbert Maier and constructed in 1931 by a team of NPS employees and California Conservation Corps members. The first rustic-style structure built in Yosemite's high country, the entrance is also the highest national park entrance in the forty-eight coterminous states, almost two miles above sea level, at 9,946 feet.[7] It would be our shelter for the night, but first we had to find a way in. The snow was up to the eaves here too, but this roof was bare, the snow blown away by the wind that screams through in winter (Tioga Pass Resort is in a more sheltered location, so the snow stays in place there). The winds scour out the upwind side of the building, as the air currents hit and then deflect around it. This leaves most of the building's south side exposed, while the others all fill in. As it happens, there is a window on the south side, so that was our (awkward, difficult) entryway to the shelter and food inside. The electricity was off but the woodstove didn't care, so we had a warm, if dark, evening, with no lanterns or other lighting available other than our small headlamps. With darkness adding to the fatigue of a long day, most of us were asleep by nine o'clock. We had an easy day to look forward to, just seven miles to Tuolumne Meadows, all downhill or flat.

In the morning, I took my cup of coffee outside and up to the rooftop, where I sat down, watching and listening to the awakening day. The station was still in shadows, the temperature in the twenties, and the air was still. In front of me was a blend of meadows and forest, the trees all whitebark and lodgepole pines. Bark beetles and blister rust have not decimated either Sierra tree like they have in the Rockies, so I enjoyed seeing a live and vibrant—if partly buried—clumping of trees. (Trees at this elevation in the Sierra enjoy abundant groundwater most years thanks to all the snow, enabling them to resist beetle attacks, and California's long and dry summers discourage the rust.[8]) With branches so flexible they can almost be bent back against themselves without snapping (at least as twigs), whitebarks tolerate the more exposed sites, though in this moment they still reflected the prevailing wind direction, pushed north by a force unseen. The whitebarks also appeared to be ever-hopeful mountaineers, climbing the slopes of Mount Dana across from me, ascending the mountainside in long strands, but withering away to krummholz after four or five hundred feet, pushed back by the same—and additional—forces unseen (wind and a climate too cold for trees). Still, the strands are resilient, and at the hour of my reverie, they disappeared into a low-hanging cloud—so who was I to say they hadn't summitted?

The momentary cloud (I could see sun down-valley on Kuna Crest) was more a mood than a shower. It was silent, hushing the wind and the land below. Snow partnered with the mood cloud in quieting the dawn land, its sound-absorption ability so good (especially when freshly fallen) that it can produce ambient soundscape readings almost at zero in some locations (0.7 decibels—the instrument's floor—at one location in Yellowstone, for example).[9] This snow was old and not so quieting, but stillness and silence were still the experience of morning, for nothing moved or spoke, except the occasional skier making the tortured climb through the window to use the pit toilet next door. Even the friendly skies were quieter than usual, the hour too early for inbound flights to the Bay Area (Yosemite is right under the area's flight paths, so high-elevation aircraft are audible 50–70 percent of the time, anywhere in the park).[10] The profound stillness and natural quiet I felt that morning would end up being one of the highlights of the trip for me, a tranquility I can sense even today—and that always came to mind when I drove through the entrance on other trips to the park.

A few hours later, we took off for the meadows. With 1,300 feet of elevation to lose in seven miles, I think most of us thought that we'd just have to point the skis downhill, perhaps with a few strides to get across a flat area or two. Jean may have been the most disappointed, for she asked for first tracks, but soon found that she could barely move without some striding. For the most part, the snow was firm—dubbed "Sierra Cement" in the downhill-skiing world, due to its high moisture content and consequent firmness—but the surface was softening in the bright sunshine. Still, I remember her expression as she passed me—one of joy. It was hard not to be happy, for we were moving, enjoying sights we ordinarily passed at forty-five miles per hour. First, the meadows near the pass, so white and seemingly pure in the way that snow makes the most unlovely landscape appear clean and crisp. Then, the Yosemite subalpine forest, the haunt of lodgepole and whitebark pine, both striving for bigness in this mountain range of giants. Rising above all were the mountains of the Sierra crest and, around a few more curves, a peek at Mount Lyell, with its remnant glacier of the same name. Finally, the Cathedral Range of arêtes right in front of us, announcing our impending arrival at Tuolumne Meadows.

Soon we were crossing the Tuolumne River, and then taking our skis off at the campground office building, used as a skier hut in winter. It's another historic building, constructed in 1936, in the rustic stone-and-log style. It would be our home for the next two nights, and because the meadows do not receive as much snow as Tioga Pass, it would be much brighter, as the windows and doors at this milder elevation don't need to be boarded up for winter. We were the only occupants for the two nights, which made it less crowded and more enjoyable.

That afternoon, some of us skied across the meadows and river to Parsons Memorial Lodge, a historic meetinghouse built by the Sierra Club in 1915 (see plate 27). In both summer and winter, the lodge appears larger than its 1,040-square-foot interior footprint would suggest. Summertime reveals the massive granite boulder pile that seems to morph organically into a building, the sides completely made up of seemingly uncut stone river cobbles, the arching doorway of sculpted stone, the roof of massive log supports under corrugated sheet metal. For us that fine winter day, it was half-buried, but with three or four feet of snow on the roof, it was still a notable presence. Built on a patch of private land near a small mineral seep, the Sierra Club donated the land and lodge in 1973 to the NPS, who today uses it to display some history exhibits and as the venue for a summer speaker series. Located on a knoll above the Tuolumne River, the lodge looks out across the meadows, with Cathedral and Unicorn Peaks rising above. Both mountains have two spires rising from rounded shoulders, giving away their glacial past. The spires stuck out above the snow, while the shoulders were smoothed by ice. Mountains that did not rise above the glaciers are today rounded granite domes like Fairview Dome, overlooking the meadows from our right (west), and Lembert Dome, the eastern sentry. Fairview rises a thousand feet from the meadows (Lembert two hundred feet less), indicating the ice was at least that deep, while the glacially rounded shoulders of both twin-spired peaks rise a thousand feet more, indicating that the two domed sentries were deeply buried (see plate 29). Parsons Lodge, then, looked out upon a multidimensional landscape of time and glaciation, spires and domes, and meadow and river, dimensions that we had to ourselves that afternoon of beauty.[11]

History is another dimension to the lodge and the river at its step, for Parsons Memorial Lodge is a national historic landmark, indicating its importance beyond local or regional scales. The building's design—in harmony with its setting, whose rigors it is clearly able to withstand—is what makes it nationally relevant, but it seems to me that there is something far more significant about this structure: the inspiration the lodge and its surroundings gave to the Sierra Club and its members, particularly in the era of its construction. Just two years previous, Congress had passed the Raker Act of 1913, authorizing San Francisco to impound the Tuolumne River in Yosemite's Hetch Hetchy Valley, some twenty miles downstream of the lodge. Conservationists like John Muir and other club members fought the dam bitterly, for Hetch Hetchy was a close second to the beauty of Yosemite Valley. Licking their wounds after their loss, the club built the lodge on the very same river, to reaffirm their commitment to Yosemite's continued preservation, and named it for Edward Taylor Parsons, who was the club's director during the Hetch Hetchy fight. Like John Muir, Parsons died the following year; both men saw their health fail, their hearts broken by the dam defeat. But, as a direct result of the Hetch Hetchy loss, the momentum to create a federal agency to oversee the country's national parks finally reached a tipping point. In 1916, Congress created the National Park Service, the primary advocate for the nation's parks, and an agency that has often partnered with the Sierra Club. Indeed, just a generation or two after the NPS's formation, Sierra Club's first executive director, David Brower, would pick up the battle against dams in the Grand Canyon. Brower made many first ascents in the Sierra, including several in Yosemite, so it's highly likely that he spent time in Parsons Lodge, absorbing the lessons it has been witness to, and then continuing the legacy of protecting our special places—without dams and reservoirs.[12]

As I skied back across the meadows, I thought about the legacy embedded in the stone and logs of Parsons Lodge, and whether we'll broaden that legacy to stave off the worst effects of climate change. As of 2014, Yosemite is 4 degrees warmer than it was in the 1920s, with virtually any measure of warmth at the extreme hot end of its historic climate (it should be noted, however, that the park's climate cooled about one degree between the

1910s and 1920s). For example, the park's annual average temperature, the average July high temperature, and the average January cold temperature were all in the 99th percentile of typical warm weather there. In fact, the only temperature metric that was not in the extreme hot end of Yosemite's climate, average winter temperature, was still in the 80th percentile of its warmest winter weather—and that was before the history-making 2012–2016 drought, which may have pushed this metric into the 99th percentile too (and which we'll discuss later in the ski trip).[13] Yosemite is heating up, and the native plants and animals are beginning to reflect the warming environment, in both subtle and obvious ways.

One of the animals demonstrating probable global warming–induced changes to its physiology and range was in hibernation right below us as we skied: the Belding's ground squirrel. Sometimes called "picket pins" for the way they stand on their hind legs when watching for predators, this hardy ground squirrel lives in the meadows and grasslands of the High Sierra. Centered in elevation on Tuolumne Meadows, the squirrel ranges well above treeline (almost to twelve thousand feet), always in areas that have a moderate cover of grass—too much and it can't watch for predators, too little and there's not enough to eat. Their hibernation occurs for more than half a year, especially for males, who head to the hibernacula (the hibernation den, usually a part of their underground burrow system) a few weeks ahead of the females; both generally head to the den around the time the grasses begin to brown.[14] Hibernation may be a good adaption to winter and late summer conditions in the Sierra, but the ground squirrel is struggling to cope with California's rapidly warming climate. Since 1966, it has disappeared from 42 percent of seventy-four sites where its presence had been recorded. All of the abandoned sites were in the animal's lower elevations, with no corresponding colonization of previously unoccupied high-elevation locations, probably because the squirrel already occupies all available high-country habitat. Warmer temperatures in both summer and winter are driving the range contraction, which translates into a loss of suitable habitat across more than eight hundred feet of elevation. Climate models indicate that by the year 2080, the Belding's ground squirrel may be pushed off the mountaintop, its once suitable habitat reduced by 50–100 percent.[15]

The Belding's ground squirrel is not alone in responding to the warming climate by moving upslope. The pika has abandoned ten of sixty-seven California locations it had been known to live in, from Mount Shasta to the southern Sierra. The abandoned sites were 1,600 feet lower than the average of the occupied ones, and modeling indicates that summer temperatures are the likely cause of abandonment. Running the same model forward in time projects that the pika may join the ground squirrel in being pushed off the mountaintop as the century progresses, disappearing from 39–88 percent of its currently occupied habitat by 2070.[16] In another study, biologists repeated some of the small mammal inventories across Yosemite done by Joseph Grinnell from 1911 to 1929. Grinnell, a professor at the University of California, Berkeley, and passionate promoter of the national parks, took periodic surveys of small mammals; he began in the San Joaquin Valley, proceeded to and then across Yosemite along Tioga Road, right through Tuolumne Meadows, and then traversed Tioga Pass and down to Lee Vining. Few parks enjoy such a methodical inventory of their small mammals; fewer still have the luxury of having the inventory done early enough to permit comparisons of environmental change over a century of their existence. In Yosemite's case, the comparison revealed that half of the twenty-eight species the biologists monitored had moved upslope, with the average movement being 1,600 feet up. High-elevation species typically pulled the lower limit of their range up, but none of them expanded upward. Two high-elevation species, the shadow chipmunk and the bushy-tailed woodrat, may already have been pushed off the mountaintop, because they were not detected in the repeat survey. Not all movement was upslope; two shrew species moved downslope, and many stayed in place. Overall, though, the general trend was upslope, even among lowland species, expanding into places that were once too cold.[17] If these critters could talk, it's clear that a good portion of them would tell us that it's getting hotter out there.

In a similar repeat survey along the entire length of mountainous California, from Mount Shasta to the southern Sierra, researchers produced comparable findings, including that when high-elevation species responded to increasing temperatures by altering their range, it was the lower boundary only. None of the high-elevation species expanded upslope; rather, they died out at the lower, warmer portions of their habitat, remaining in place elsewhere.[18] Both groups of researchers noticed this range contraction, but neither investigated it further. It could be that the hardy species that inhabit the cold and snowy elevations near and above treeline in the Sierra have already occupied all suitable habitats up to the mountaintops. Well adapted to the forbidding climate at the highest elevations, the pika and its consorts would probably not be deterred by temperatures just a few degrees colder, which means, further, that little prevents them from colonizing all available pockets of habitat up to the mountaintops. If this is the case, then we may need to adjust our projections regarding the availability of unoccupied habitat above a high-elevation species' current range. At least in the Sierra, the land above treeline is little more than rock, water (liquid and frozen), and sky. High-elevation small mammals may already be at their upper elevation limits, so climate change may already be forcing species off the mountaintops. Indeed, the next time we run the small mammal surveys, we may well find that another one or two species have joined the shadow chipmunk and bushy-tailed woodrat in becoming regionally or nationally extinct.

With birds, the situation is more complex, though there has still been significant range shifting since early 1900s. Like he did with small mammals, Grinnell surveyed birds along periodic transects to and through Yosemite (and Lassen Volcanic National Park and Sequoia National Park), and like the mammal resurveys, the bird surveys have been repeated by a team of researchers more recently. The team found that all but five of the ninety-nine species they encountered had moved up, down, or both directions, and the five that did not move were low-elevation species that are not sensitive to human presence and able to utilize neighborhood food sources. Of the ninety-four species that experienced range shifts, 51 percent moved upslope in response to the warming climate, but the movements of the remaining species were better explained as a response to increasing precipitation, which commonly drove those birds down in elevation (wet years in the Sierra appear to be getting wetter, as we will discuss more below). To resolve the conflict that arises for a bird being pulled up by rising temperatures and down by increasing precipitation, the research team devised a "nearest neighbor" model that looked for nearby areas that offered the bird refuge from both aspects of its changing climate, such as a lower-elevation canyon whose north-facing slopes offered shelter from hotter weather. This combined approach explained a much higher portion—82 percent—of the changes in bird ranges over the last century. The birds most likely to shift their ranges were those with small clutches, who defended all-purpose territories (for example, those for breeding and foraging), and who were year-round residents—in other words, those with the most to lose. Overall, most birds are responding to the changing climate, with over half moving upslope, much like the small mammals. When avian responses to increasing precipitation are also considered, a hefty majority of the Sierra's resident birds are responding to climate change.[19] Indeed, all flora and fauna are sensitive to climate, but their vulnerability depends on how they adapt to changing conditions.

Butterflies, in fact, are another winged animal in Yosemite responding to climate change. Analyzing thirty-five years of data (beginning in 1972 and ending in 2007) at ten sites stretching from near sea level to 9,100 feet above, and including 159 species, a team of researchers found that the richness of species diversity declined at half the sites, most notably at the low-elevation sites where habitat destruction is greatest. But at the highest location, the opposite was true: more species were present in later surveys (except for some alpine specialists that appear to be declining), and most of them increased in abundance as well. In the higher-elevation sites, more generally, the team also found clear evidence that many butterflies were moving upslope, consistent with a warming climate.[20] Butterflies, birds, and small mammals are speaking with one voice: the climate is warming, and the majority of their movements in the last century are probably a response to increasing temperatures and/or precipitation.

Back at the cabin, I joined the group for a hearty dinner. Afterward, I went out to the snow bench I had shoveled out earlier in the day, on the banks of the nascent Tuolumne River, to watch the sunset. No one joined me, perhaps because it sounded cold to them. Having just left the frigid plateaus of Yellowstone eighteen months earlier, I still had enough antifreeze in my bloodstream to keep me warm in subzero weather, so the relative balm of the High Sierra in April was practically T-shirt-and-shorts weather to me. I spent an hour on my unconventional bench, sipping tea (with a sweetener of Southern Comfort) and watching the colors of the eve meld with the night's, blues becoming charcoal, then black. At my feet gurgled the Tuolumne, there just the size of a creek, barely a hint of what it would be in a few weeks, once all that snow began to melt. On the far bank, just a few feet away from me, was two feet of snow. At river's edge, it formed billowy pillows here and layered walls there, the walls formed where the snowbank had calved into the river below. Above the pillows and walls was a meadow, its grasses and willows just a daydream that April eve, buried by a blanket of snow. A hundred yards away, the ground rose out of the floodplain, the drier conditions there suitable for lodgepole pines—a whole darkened forest of them. Both meadow and forest were well on their way to the colors and shades of night, the nearly black trees edging ahead of the twilight-blue meadow to their final destination, the ink of night. Above the forest rose the Sierra Crest, the bald, snow-dusted mountaintops defining Yosemite's eastern boundary. The highest summits held on to glimpses of light from the sun, now gone below the horizon, shoulders of brightness that seemingly glowed in a landscape of soft purples, blues, and grays. Soon the bright spots, so subtle a contrast to the darkening land below them, winked out, joining the land in its timeless move toward nightfall. With the river's gurgling helping to mask the high-altitude jet noise, it was a setting—both for the eyes and ears—of natural tranquility, as rich an encounter with nature's serenity and beauty as coffee on the rooftop at Tioga Pass had been that morning, twin bookends to a delightful day.

The next day was the layover day, ours to explore the meadows or to hang out at the cabin, or both. Excited to be in Tuolumne for the day, I took two ski tours (as any outing on cross-country skis can be called), a morning tour to Elizabeth Lake and an afternoon one to the top of Lembert Dome, or close to it. Terry and Bruce joined me for the tour to Elizabeth Lake, a classic Sierra tarn nestled at the base of Unicorn Peak. The lake is only a couple of miles from the meadows, so it didn't take us long to get there, even with our climbing skins slowing us down (skins are strips of knapped fabric that adhere to the bottom of skis and allow one to climb). By this time of the snow season, the snow has gone through multiple freeze-thaw cycles, firming it up enough to walk on without sinking in much. Skis are much faster, though, so they remain the preferred means of winter wilderness travel. After an hour of uphill skiing, we broke out of forest and into a meadow that led us the final third of a mile to the lake. The lake was discernible from the whiteness of meadow only by virtue of being flat, its surface frozen and equally brilliant. Unicorn Peak rose straight up from the far shore, its slopes marked by avalanche scars, those that slide frequently distinguishable by the sapling pines growing in the lesser-used slide paths. Those saplings will confront the reality of their ill-fated home the next time conditions become ripe for avalanches in the seldom-used slide paths (which occur maybe once a century). Outside of the avalanche scars rose a smattering of trees, shrinking in stature the closer to the alpine they grew. The mountain's spires were out of view, our lake too close to the mountainside to permit a sighting, though we could see the shoulders just below the spires. We enjoyed the beauty at Elizabeth Lake—and the power inherent in this landscape of snow and spires—for a half hour or so, then made ready for the return: the downhill thrills. Taking the skins off, we pointed the boards of speed downhill and commenced our Hail Mary run back to the cabin. Zooming through the forest and avoiding collisions with the arboreal residents was exciting, and we made it down in a fraction of the time it took to get up.

The forest we whizzed through, like most Sierra forests, is experiencing some subtle changes, sobering indications that climate change is likely altering these magnificent spaces in ways that soon may not be so subtle. Much of this begins with the fact that Yosemite's climate is already at the dry end of what the trees there can tolerate. Since the early 1700s, the annual water deficit has grown by 5 percent (water deficit is a measure of drought stress), due mostly to that 4-degree rise in temperature.[21] This means that trees in Yosemite were already contending with a drying climate before the effects of anthropogenic climate change began to be felt. This additive effect may explain why the background rate of tree mortality in California has doubled since 1980 (from 0.9 percent to 1.8 percent annually), across all ages, elevations, species (though pines and firs lead the dying trees), life histories, and fire-return intervals.[22] The biggest trees in each species have been hit hardest, beginning in the 1930s.[23] Since 1908, the maximum elevation at which a fire will burn in Yosemite has risen over 1,600 feet. While fires at such heights constitute just 2 percent of all fires, the probability of fire above about 9,800 feet has quadrupled since 1989.[24] The increased burning is probably related to a surge in growth rates in the latter half of the twentieth century in the region's highest-elevation forests, as seen in the growth rings of treeline species in the Tioga Pass area and the White Mountains just east of the Sierra Nevada. Recall that tree growth at the upper treeline is limited by temperature, not precipitation, so the trees there are responding to a warmer and longer growing season.[25] Throughout the twentieth century, the highest-elevation trees in Yosemite have also been invading formerly persistent snowfields and adding branch growth, two more indications that the upper treeline may be rising.[26] In general, conditions appear to be drying in the lower elevations—where moisture limits tree growth—and mid-elevations, and warming across all, so the forests are responding accordingly. It would not be long after this ski trip ended before some Yosemite forests began to respond more obviously to the warming and drying climate—as we would see signs of in a day or two.

Once we'd eaten lunch, the whole group of us took off for the climb to Lembert Dome. We took the trail on the south side, which steeply ascends three hundred feet, so the skins came in handy again. Once we had gained the ridge leading to the summit, we turned left and continued to where the forest ended and granite began, another two hundred feet up. There, unfortunately, we were turned around by a steeply pitched snowbank that we would have to traverse to gain the summit. Some of the group headed home, but before Bruce and I did the same, I suggested we check out an opening I could see through the trees, just north of the ridgeline. It turned out to be a shoulder of Lembert Dome that was not impeded by a sketchy snowbank. It was not quite as high as the dome itself was, but it offered much the same view. In every direction was a glacial landscape in winter garb, from the mountaintops (where the glaciers began) lining the entire eastern horizon from north to south, to the domes, spires, and meadows that they molded to the west. The field of arêtes bounded the meadows to the southwest, with Echo Peak, another arête, adding its twin spires to those of Unicorn and Cathedral Peaks. Moving to the northwest, Mount Hoffman and Tuolumne Peak appeared, mountains in the center of Yosemite that pushed the accumulating ice over the meadows. There, the accumulated ice began to move northwest and then down the Grand Canyon of the Tuolumne River. And in the foreground were the whiteness and greenness of the meadows and surrounding forests, bounded by a collection of granite domes, large and small. Roches moutonnées the domes were, flocks of stone sheep that have all been rounded, shaped, and smoothed on the upriver side, but steepened on their downriver side, both by rivers of ice.[27] The ice caps and glaciers may be long gone, but they're well remembered by this landscape of geometric glaciology.

Were we to return to this dome in summer (as I often did in my Yosemite tenure), we'd see that the summertime warmth had not erased whiteness from the landscape completely. Ten or twelve miles to the south, on the slopes of 13,114-foot Mount Lyell, Yosemite's highest peak, we would see Lyell Glacier, bright white in the California sunshine (see plate 30). One of five remnant glaciers in or very near the park, Lyell Glacier is the most familiar, visible from the Tioga Road in at least one place; it's also the second largest of the Sierra's fourteen glaciers (Palisade Glacier in Kings Canyon is the largest). That any glacier survives in the relatively balmy Sierra Nevada amazes me, but their presence is a testament to the prodigious amount of snow the range receives.

John Muir himself was the first to study Lyell Glacier and its nearby cousin, Maclure Glacier, pounding a series of hand-hewn stakes in a straight line across both glaciers in August 1872. Returning forty-six days later, Muir found his stakes were no longer in a straight line; rather, the two end stakes were still in place, but now the remainder were bowed, with those near the glacier's middle displaced the most, just as they would be if placed in a liquid river. Measuring the maximum displacement in the center (using a plumb line he fashioned from a rock and a horse's tail hair to estimate the previous position of his stakes), Muir found that

Maclure Glacier had moved forty-seven inches, or an inch a day (Lyell Glacier was similar). Muir thus began a tradition of monitoring the two glaciers that is approaching its sesquicentennial, which is probably the longest-running glacial research and monitoring effort in the country. Throughout the last 150 years, park rangers and researchers have trekked to the glaciers to record their positions and, in the case of Lyell Glacier, take pictures of it from a fixed location (topography does not offer a convenient photographic vantage of Maclure Glacier). They have also found one of the stakes Muir used, transported to the glacier's terminus by moving ice, along with a mummified bighorn sheep. There are gaps in the record, but this is an exceptional history of a notable Sierra Nevada glacier.[28]

That history may soon be coming to an end, for the research and information gathered since Muir pounded his stakes into the two glaciers reveals a long march of retreat. Like most Sierra glaciers, these two have lost significant area: as of 2004, Maclure has lost almost half its area (47 percent) and Lyell has split into two halves, the eastern one losing most of its spatial coverage (78 percent) and the western the least of the three (40 percent). More recently, Yosemite geologist Greg Stock repeated Muir's study, pounding more contemporary stakes into the two glaciers, on the same exact dates that Muir did his work. True to historic precedent, Stock returned forty-six days later, and was blown away by his findings at Maclure Glacier: it had moved forty-seven inches in the elapsed time. Even though the glacier had shrunk by then to less than half the size it was in Muir's day, it was still moving at the exact same rate, an inch a day. On Lyell Glacier, Stock's results were more sobering, for his stakes had not moved—and that was on the glacier's western lobe, the eastern one too far gone to bother measuring. By definition, glaciers are moving bodies of ice, so it appears that the glacier stagnated sometime between the 1930s, when its velocity was last measured, and 2012, when Stock measured it. Since then, record heat and drought have exposed bedrock in the middle of the western lobe, allowing him to estimate the remaining ice thickness—it's only 16.4–32.8 feet thick, and the former glacier lost a sizeable portion of that in the recent drought, so Stock estimates it could be gone completely in a decade or so.[29]

I first learned of Lyell Glacier's demise when I was still working in Yosemite and Stock gave a presentation to his coworkers, probably in early 2013. He had arranged for the park archivist to bring along Muir's stake, which had been added to the park's museum collection after it was found at the glacier's foot. Seeing and touching it was a novelty, but other impressions from the presentation have stuck with me more, all of them variations on a theme of loss—self-inflicted loss. The first, and perhaps most basic, loss is that the end of Lyell Glacier will impact everything downstream, especially in late summer and in drought years. It won't just be the plants and animals that will be affected, but also the people of San Francisco, who depend on the Tuolumne River for their drinking water. Another loss is that a beloved natural landmark had been pronounced dead. It was akin to what the residents of New Hampshire felt when the stone face known as Old Man of the Mountain collapsed in April 2003. The rock outcrop had become so much a symbol of the state that people left flowers on the site after it fell.[30] Lyell Glacier was certainly not as widely known as the Old Man was, but it was well loved in the Tuolumne high country. A third loss is the impending end of Yosemite's long-running glacier research and monitoring effort, and the insights it has given us. As Stock explained in his presentation, Maclure Glacier is probably also in its death throes, soon to become so thin it will not have enough mass for gravity to move downhill. Within a decade, then, Yosemite's glaciers will cease to exist, and the research effort will draw to a close, with no more conclusions to provide. The final loss is that we will lose a sensitive indicator of environmental change. For more than 150 years, Lyell and Maclure Glaciers have been bellwethers of our changing climate, as have all glaciers worldwide. As they melt away into oblivion, we force the world into uncharted climate terrain, with fewer and fewer bellwethers for perspective. In at least these four ways, the story of Yosemite's glaciers tells us that their loss is not just biological and physical but profoundly social as well. Indeed, their most important value may well be their symbolism as indicators of a healthy climate and a pristine environment.

In Missouri, once my legs buckled underneath me, I had little choice but to begin life in a wheelchair. It was definitely the lesser of two evils, for the other choice was to fall helplessly to the floor, potentially taking out one or both of my caregivers with me. The wheelchair offered security, so the transition to it was straightforward, if not welcome. My folks got to be adept at using a patient lift to transfer me to and from bed, the wheelchair, my recliner, and my desk chair, using a patient sling with hooks that attached to the lift. Putting me in different chairs helped prevent the bedsores I would have gotten from sitting in one chair all day, every day. My leg muscles were still strong, so it seemed like I should have been able to do more, but a nurse reminded me that my brain could no longer reach all my leg muscles. As time wore on, they continued to weaken, soon making it impossible for me to push myself up in the chair, and eventually weakening so badly I could not fully extend my legs in bed. My folks grew accustomed to me asking them to "hold my knees please," which referred to our technique for moving me from a slouched position to an upright one: it consisted of leaning the wheelchair seat back (to get gravity on or our side), and then simultaneously pulling me up from my armpits (Dad) and pushing up against my knees (Mom). I actually still have residual control over my legs, though if a miracle cure were discovered tomorrow, I question whether I would have enough muscle strength to stand. Whatever that speculation might be, my reality is that, with my body unable to maintain its position in the chair, I can't withstand more than an hour of being bounced around in a moving vehicle. The thrill of hiking and exploring new and beautiful places in the High Sierra, the physical satisfaction from pedaling my bike on Forest Service dirt roads just outside Yosemite, and the coolness of swimming in the Merced River seem increasingly like dreams—good ones to be sure, and ones that always inspire—from a separate life, one that ended long ago and far away. Indeed, my travels and physical activities are over, my body increasingly a prison cell, not only for me, but, in some ways, also for my caregivers.

As it went for my legs, my remaining strength continued eroding, with two possibly related experiences marking the decline. The first was that, with no remaining neurons left to destroy, I began losing complete control of entire parts of my body. First were my arms and hands, whose control had been faltering for some time. The last things to go on both arms were two fingers on each hand, and the pattern of their demise was revealing. For several months, I had been able to move the four fingers a little, my movements resembling the small, twitching fasciculations that are a hallmark of ALS—except that these were intentional, not involuntary. One by one, though, each of the four fingers began to twitch involuntarily for a day or two, after which they stopped any kind of movement. The day or two of involuntary twitching was when the connections between the brain and the motor neurons were being destroyed, followed by paralysis. Those fingers have not moved since, either of their own accord or of mine, with a few involuntary exceptions that probably destroyed a leftover neuronal connection or two that escaped the earlier destruction. Since then, for all the good my arms have done me, I've had a newfound respect for amputees.

The other experience of being wheelchair bound is that the disease progression seems to have slowed, at least at times and in some parts of my body. This has probably been most evident in my breathing ability, which has seen long periods of relative stasis, as though it were on a low plateau, before dropping to the next lower plateau. Perhaps this slowing is due to the fact that most of my body's motor neuronal connections are already gone, so the agent of destruction—whatever it may be—is left searching for the few intact connections that remain. It could also be a matter of perspective, since now there are fewer benchmarks to measure the progression against. In the early stages, the decline in physical abilities was dramatic and easy to see: I couldn't keep my balance on cross-country skis near Yellowstone, or on the rocky part of the trail near Yosemite, or, finally, getting into the raft on the Grand Canyon river trip. Then, I could compare my increasing disability to what I had been able to do a few months, or even just a few weeks, before the pivotal events—but now, I have few reference points that offer useful comparisons. Take my legs, for example: being wheelchair bound, I have no use for them, so I don't engage in any activities that demand leg use. As a result, I cannot say that I could do an action two weeks ago that I can't do now. Another time the rate seemed suppressed was at the disease's onset, because the initial symptoms developed slowly. Whatever the case may be, it's almost a given that I wish the reduction in progression rate had occurred in the early stages of my ALS. Either way, the insidious, unrelenting progression defines the lives of ALS sufferers.

When I returned to the cabin in Tuolumne Meadows, I had a surprise: in taking off my skis, I saw that one of them had cracked in half, right under the binding. The only thing holding the two halves together was the laminate layer glued to the bottom of the ski that has the "fish scales" (akin to knapped fabric) that provide for forward movement on flat and gently inclined slopes. The thin laminate would probably not be strong enough to hold the two halves together for the twenty miles remaining in our trip, so I had a problem. I brought it to Terry's attention, hoping our fearless and well-prepared group leader would have the rudiments of a repair kit in his pack. He did, so the engineers (and wannabe engineers) in the group got to work on a fix that would get me to Yosemite Valley. From somewhere, we obtained a piece of one-eighth-inch plastic, cut a smaller piece of it to size, removed the binding, molded and superglued the plastic piece to the ski, and then put the binding back in place to hold the fix—hopefully. Plastic had clearly saved the day—twice, at that (the laminate layer and the repair layer)—and it would get me out of the winter wilderness, without further ado.

The next day, it was time to move on and continue our westward trek on the snow-covered Tioga Road. The NPS manages the entire Tioga Road corridor as wilderness in winter, so even though we'd be skiing on a road that's ordinarily busy when most people encounter it, we would have it to ourselves. I began the morning with coffee on the snow bench, a rich solitary experience once again. I noticed additional places where the river had broken out of its winter shackles, iced over no more, the water's gurgling perhaps a decibel or two louder. The chickadees murmured quietly in the pines behind me, going about their business without evident concern about the primate nearby. Between the gurgles and murmurs and the early hour (the sun hadn't yet broken the horizon), the sound of high-altitude flights was somewhat reduced. Solitude, tranquility, and beauty were the defining elements of my times on the snow bench and rooftop, elements that also characterized the entire trip, as we experienced that day. We followed the road west along the edge of the meadows, pausing to look back as we left them at Pothole Dome. Then we pushed on, passing through a land of domes, large and small, with names that spoke of Yosemite's past. For example, there was Medlicott Dome, named for Henry P. Medlicott, one of the surveyors who laid out the Great Sierra Wagon Road across Yosemite, the predecessor of the road we skied on. Another was Pywiack Dome, which translates to "glistening rock" in the language of the Ahwahneechee, who once lived in Yosemite Valley.[31] The glistening rock is glacial polish, outcrops of granite that glaciers smoothed so well they glint in sunshine without being wet. Striations—linear scratches in the Pywiack—were gouged by rocks embedded in the ice, giving away the glacier's direction of travel. Glacial polish is common in Yosemite, especially in the Tenaya Lake area, whose frozen shore we soon saw. We were hungry and a bare spot on the Pywiack soon beckoned, so we stopped for lunch. The lake, normally thronged with summer visitors, was ours alone, nature's silence and beauty substituting for children's squeals of delight as they touch its cool, clear water.

I was one such swimmer on a summer day in my Yosemite tenure, but instead of squealing, I did the backstroke out to one of several tree stumps in the lake, stumps that break the lake surface farther out from shore. It was not a long swim, just five or ten minutes, and when I reached the stump, I pulled myself onto it and looked down, trying to see its bottom. The tree trunk, two feet in diameter at the lake surface, disappeared into the watery depths, well beyond my sight, probably twenty or thirty feet down. Few trees anywhere tolerate such inundation, so what are they doing there? How did they grow in water so deep? The answer lies in their very growth rings, a key to climates past. The tree stumps all date to the Medieval Warm Period, a time of globally elevated temperatures that fostered the colonization of Greenland, extending from around AD 1100 to 1375. With the exception of about thirty years in this period, California was in a megadrought that lasted one or two centuries, drying Tenaya Lake up completely or to a small shadow of its current self. Trees grew where waters now lap, and when the climate turned wetter, the lake filled up, entombing the trees that it then killed for centuries, its cold water perfect for preserving organic materials. These were droughts that reduced precipitation anywhere from 15–20 percent of the twentieth-century average to as much as 80–90 percent of it, for so long that it became the norm.[32] It was severe, enduring drought such as our society has never seen; indeed, tree-ring research indicates that the twentieth century is one of the two wettest in the last millennium in the Sierra.[33] It was an austere climate to which we might be forcing our return—and one that we got a preview of, beginning just a few months after my relaxing April lunch.

The next winter was as dry as this one was wet, bringing just 45 percent of average snowfall, as measured in SWE on April 1 of each year.[34] California's Mediterranean climate delivers 95 percent of annual precipitation between October and May, with two-thirds of it falling in the winter months, December through March. The bulk of the moisture comes in a handful of events that cause a dominant ridge of high pressure over the northeast Pacific Ocean to break down, opening the door to bucketloads of Pacific moisture. These events make or break the state's "water year" (October 1 to the following September 30, and usually tagged to the year of the April 1 SWE, as in "water year 2012" for the 2011–2012 season), with summer thunderstorms contributing very little to the available water, with the water that does come from storms usually occurring just over the High Sierra.[35] Such low snowfall is not unusual in Yosemite and the Sierra: it has happened four times in the previous forty years, two of those times back-to-back in the water years 1976 and 1977, during a drought previously considered extreme. The following winter, water year 2013, brought 53 percent of average snowfall, a modest improvement, but not enough to change the direction of the unfolding drought. Still, this drought was nothing for the history books—but that was about to change.

The next winter was not only abysmal in terms of precipitation, bringing just 31 percent of average snowfall, but it was also the warmest on record.[36] Coupling drought with warmth has happened in California's past, but only 42 percent of the time *both* occur. Since the mid-1990s, though, that percentage has more than doubled, meaning that 91 percent of drought years are also warm.[37] Three-year-long droughts have also happened in the state's past, but the record-high temperatures in water year 2014, combined with three years of below-average precipitation, produced the most severe drought in the last 1,200 years, according to two researchers examining the tree ring records. No other single year and no other three-year sequence of dry years produced soil moisture levels as low as those in 2014. Although precipitation was at the third-lowest rate in California history, it was within the range of natural variability, leaving the increasing temperatures as the likely cause of the critically low soil moisture.[38] And it truly was warm; I had many lunches outside on my deck that winter—often without a shirt—and locals started calling it "June-uary." More significantly, I remember waiting, and waiting, and waiting for a big winter storm to come through, but other than a particularly cold storm system that dropped snow down to two thousand feet, only one or two smaller storms moved through. By spring, a couple of sapling incense cedars in my yard were dying, deaths that were an ominous indication of what was soon to happen, though no one knew it at the time.

The next winter, the unthinkable happened: *no* snow fell—anywhere in California. The High Sierra did get a few dustings (and some locations received rain), but no place in the country's third-largest state got anything more than a pittance of snow. The northeast Pacific high-pressure ridge never broke down, earning itself the nickname "ridiculously resilient ridge." Yosemite's April 1 SWE was just 4 percent of average, by far the lowest ever measured in the instrument record and the lowest in the tree-ring records in 500 years. June-uary continued as well, with water year 2015 even warmer than the record-setting year before.[39] Elevations as high as 9,200 feet had no snow on the ground on April 1, and the only sites that did not set a record-low snowpack were the highest-elevation snow-monitoring stations. The stark conditions were not limited to California, for the entire Pacific coast was also affected, as we saw in Olympic. Eighty-one percent of 562 snowpack-monitoring sites in the three Pacific states (plus a few sites in western Idaho and Nevada) measured their record-low snowpacks on April 1, 2015, with 111 recording no snow for the first time ever. The return interval on such a dry winter is 400 to 1,000 years (meaning that one would not expect another winter that dry for another four to ten centuries)—and this doesn't take into consideration previous drought years.[40] When we factor them into the estimation of return intervals, we find that the 2012–2014 drought was a 1-in-10,000-years event. Including 2015 in the calculation is impossible to do, for it is completely without precedent in the tree-ring record, not occurring once in the last 1,200 years.[41] Truly, it is hard to overstate the severity and unprecedented nature of this drought.

It's probably not a surprise that our fossil fuel emissions helped push the drought into the record books. First of all, the Pacific coast states are already 1.8 degrees warmer than they were a hundred years ago, which was reflected in the record-breaking warmth in 2014 and 2015.[42] Second, several teams of researchers have used climate models to determine the human contribution to this drought's severity, running them until they hindcast the California climate correctly, and then running them through the drought without our greenhouse gas emissions, with the difference attributable to us. One team found that human-induced warming has reduced snowpack levels by 25 percent, with low and middle elevations seeing a greater decline, 26 percent and 43 percent, respectively. With mid-elevations (about five thousand to eight thousand feet) constituting 60 percent of the Sierra's area, such a reduction equates to a huge loss of meltwater, especially in drought years. This team also found that, above eight thousand feet, anthropogenic warming has reduced snowpacks by 10 percent, probably by shortening winter.[43] Another team found that, with oceans absorbing much of the planet's increasing warmth, higher sea surface temperatures amplified the high-pressure ridge in the North Pacific, shunting winter storm systems away from California. Up to one-third of the variance in California's recent precipitation is due to this enhancement of the instrumental high-pressure ridge—and, thus, to our emissions.[44] Although still another team downplayed the influence of the high-pressure ridge on California's drought, it was still a factor there, just more influential in Oregon and Washington. Also, because the highest-elevation snowpack-monitoring sites did not see record-low snowfall in the 2015 water year (while lower ones did), this team nonetheless concluded that anthropogenic warming exacerbated the Sierra Nevada drought.[45] (Note that amplifying the northeast Pacific high-pressure ridge does not necessarily come at the expense of its breakdowns in very wet years, so our emissions may be amplifying the wide variability already present in California's climate from year to year, though this has not been verified.[46])

Finally, two additional teams used novel approaches to discern the human influence on the drought. One paired a hydrologic model with a risk-assessment framework to estimate the probability that the 2012–2014 California drought would have been less severe without anthropogenic warming. Had the temperatures in water year 2014 resembled those of the foregoing century—which themselves were on the increase—there is an 86 percent chance that the drought and its accompanying soil moisture levels would have been less severe.[47] The other team used the Palmer Drought Severity Index (a widely used measure of drought intensity), to get at the question. Their results echoed those of the other teams, finding that 5–18 percent of the drought anomaly in 2014, and 8–27 percent of the previous three years' anomaly (2012–2014, as a group), were attributable to human-caused global warming.[48] In sum, five different teams of researchers, using five different methodologies, all arrived at the same conclusion: human-caused atmospheric warming exacerbated an already extreme natural event, pushing it into the realm of a never-before-seen phenomenon. We reduced snowfall by at least 8 percent, and as much as a third or even more.[49] This may not seem like a significant amount of meltwater, but it might have been enough to keep the drought from exceeding the severity of the infamous 1976–1977 drought. It's safe to say that anthropogenic warming is already forcing the California climate into the world of science fiction, one in which humans, if present trends continue, may well come to play as big a role as nature itself in determining how far the climatic extremes go.

As months turned into years in Missouri, my swallowing and speaking abilities continued to erode. Knowing the disease progression would only make eating more difficult, my neurologist recommended I have a feeding port installed in my stomach. This is nothing more than a short tube inserted through the wall of the stomach, which provides an alternative, easy, and noninvasive way of giving someone liquid nutrition. I could see where my eating situation was headed, so I had one installed about a year after I arrived in St. Louis. Another year later, and after a typically tiring ALS clinic, I used it for the first time (I had it put in while my breathing was still good enough to be put under for the procedure). I had choked several times on my dinner that night, so after a notably long larynx spasm, I requested that the rest of my dinner be the formula given to people in my situation, known by its retail name of Jevity. With the self-imposed floodgates against supplemental nutrition opened, the long-delayed acceptance of my dietary fate—100 percent liquid, delivered through the feeding port—gradually started to become a reality. Increasingly, I ate "for pleasure," just to enjoy the taste of food, getting most of my actual nutrition from Jevity. Landmark moments included my last meal out, at a winery in Hermann, Missouri (quiche, a texture and consistency I could eat, carefully); my last ice cream at Ted Drewes, a local ice creamery (a raspberry concrete, the last texture I could handle safely); and the last two foods I enjoyed at home, coffee and peaches, both favorites of mine (the coffee thickened to a paste, the peaches pureed to the same). By this point, my tongue could do little more than remain in place, so the tiny bites I took of either paste accumulated in my mouth until my folks suctioned up the mess. Once the fresh peach season ended, I knew it was time to seal my fate: no more food by mouth, even for pleasure. Much like a death row prisoner the evening of his execution, my final meal was over.

Given the tongue's importance in both swallowing and speaking, it's not a surprise that the gift of speech was also leaving me. Slowly, gradually, I came to rely exclusively on the speech devices to communicate anything at all. I went through a series of them as opportunities became available to upgrade without shelling out $14,000, with the models generally improving in terms of computer creation and processing speed. My parents, brothers, and a good friend could often discern what I needed in situations where the speech devices would not work (like outside, where there is too much light). Even then, there were times when I just gave up if it was something I could live without, or we resorted to using an alphabet chart to spell out a key word that would enable them to figure out my need. Many people who knew me before ALS have told me they miss our conversations. Of course, I do too. Notable landmarks along the way to becoming mute included giving up my recliner because, with the speech device mounted to my wheelchair, being in my recliner meant I had no way to communicate; arriving to the point where I can only say three words, "No" and "I am" (hot or cold, with my folks guessing which one), all in limited situations, and only to my parents; and passing people on walks in public places but being unable to return their greeting (which is ongoing). If not for the speech device, though, I would be feeling increasingly isolated and alone, unable to communicate even the most basic needs. It's easy to see how, before eye-tracking technology became available, ALS diagnoses would result in earlier deaths than they do today, with sufferers unable to tell anyone what was happening and what they needed.

I use a similar device on my computer, without the speech component. By tracking my eyes about an on-screen keyboard, I'm able to use Microsoft Word (the computer screen is larger than the speech device's, otherwise I would write using the device). Being able to write has allowed me to savor some good news, such as the publication of my third and fourth books. Both of them allowed me not only to relive the hikes and other adventures in Yellowstone—much the same as this book has done with other parks—but also to reflect on their meaning. Had I never gotten cursed with ALS, I might not have done such reflecting, at least until I was older, nor might I have distilled them in writing, and certainly not this soon. This part is just speculation, but the satisfaction and enjoyment of seeing the two books published is definitely not conjectural. In reminiscing, it's also natural to think of the threats to continued national park preservation; doing so became the nucleus of this book, climate change being the single greatest threat to most of our national parks. This is not to say that I would ever willingly accept the disease, but rather to state that it—or any other disease—may open other doors. Also, it's important for anyone in a situation like mine to have a way to remain productive and engaged, with goals to strive for. Naturally, these will differ for everyone, but persons in a wheelchair, by virtue of having a debilitating, terminal illness, enjoy a bully pulpit of sorts for speaking out about issues of personal and societal concern. Essentially, if a person with a shortened lease on life still speaks forth about an issue of public concern, shouldn't the rest of us take notice?

Back at Yosemite's Tenaya Lake, we finished lunch and resumed skiing. We considered skiing across the lake (which is more a more direct route than the road), but the ice at the lake edges was discolored in places, indicating it was beginning to thaw. Playing it safe, we stayed on the road, skirting the lake and enjoying the solitude of a place usually thronged with visitors and boisterous children. Partway up the climb to Olmsted Point on the other side of the lake, we hung a right onto the old Tioga Road, which avoids an avalanche zone and is a more direct route to the last cabin of our trip. That particular avalanche zone has the notoriety of spawning a snowslide that killed a park employee. In June 1995, a large slab of snow broke loose, sliding down the naked granite slopes above the road and hitting the bulldozer that Barry Hance was operating in the park's annual effort to open the Tioga Road. The dozer was flipped on its side and buried by heavy, wet snow. By the time his coworkers exhumed the cab, Hance had succumbed. The plowing operation was halted that year, and ever since, park managers wait until the risk of avalanches subsides, both there and in the other avalanche zones through which the road passes.[50] Hance was a well-liked, longtime employee when he died, at the age of just forty-three. In his honor, park managers bestow an award every year to the employee who best exemplifies the attributes people admired in Hance. Fittingly, it's Yosemite's highest personnel award.

With plenty of daylight to spare, we found the Snow Flat Cabin in the forest near the May Lake trailhead. Only the cabin's roof crown rose above the snow, this area being one of Yosemite's snow belts (as the cabin's name suggests). As winter storms barrel in from the Pacific and move up and over the Sierra, they cool, making them drop the moisture they're carrying. The foothills and elevations up to about 5,000 feet receive most of their moisture as rain, while elevations above about 6,000 feet receive mostly snow (with the intervening 1,000 feet a transition zone). Moving across the Sierra from the southwest, the storms drop most of their moisture on the first mountain or upland they hit, such as the Cathedral Range or, further south in Yosemite, the Clark Range (both of them subranges of the Sierra Nevada). This explains why Tuolumne Meadows had less snow than Snow Flat did, even though they are both about 8,600 feet above sea level: Snow Flat is the first upland the storms encounter, while Tuolumne Meadows is in the rain shadow of the Cathedral Range, which intercepts the moisture headed northeastward, effectively wringing the clouds of their heavier moisture. While Snow Flat will have forty-four inches of SWE on April 1 of an average year, Tuolumne Meadows will not even have half that (twenty-one inches). The same principle is also evident at Tioga Pass, which will have just an inch more SWE on April 1, twenty-two inches.[51] The colder temperatures at Tioga, 1,300 feet higher than the Meadows and, therefore, 5 or 6 degrees colder, may mask some of the diminishing snowfall by preventing the snow from melting between storm systems. These differences may seem semantic, but they help explain why the forest around Snow Flat is more diverse than that in Tuolumne Meadows, because the former is wetter, friendlier to more tree species.

The Snow Flat Cabin is mainly used by snow-survey teams, who measure the snowpack there several times a winter. Once a month from January to May, two or three park employees trek on cross-country skis to several fixed locations to measure the amount of snow on the ground. Carrying metal tubes and a scale, they pound the tubes into the snow down to the ground, extract a column of snow, weigh it, and record the measurement. I was fortunate to be a team member on a four-day trip to five different "snow courses," as they're called, north of here. The year after I took this Trans-Sierra ski trip, I skied from the Hetch Hetchy Reservoir to Grace Meadow in Yosemite's far north with Mark Fincher and Jesse Chakrin, two other Yosemite employees. The snow courses were scattered along the route, ranging from 6,500 feet in elevation to 8,900 feet. Like many such snow-survey trips, ours offered both opportunities for levity (often provided by the snow itself) and serious reflection. In this case, the levity came at my proverbial expense, as I had trouble stepping across a small open stream on the way in—and again on the return. To cross a small stream or other minor obstacle with skis on, the skier begins by standing next to the rivulet, skis parallel to the water. Next, the skier steps across it with the streamside ski, taking care to leave enough room for the other ski on the far bank. The final move is to bring the other foot across, followed by resuming forward travel. I had done this move hundreds of times successfully in Yellowstone, but the first attempt at this one found my trailing foot and ski dipping into the creek when I slipped on the far side. On our way out, I opted to step across without my pack, and then catch the pack, tossed across by a teammate. Part A went according to plan, but part B, the toss, did not—my pack landed in the stream. I retrieved it just a few feet downstream and found the contents a little damp but otherwise functional (much like my foot on the first attempt at humoring my teammates). Since then, Fincher, who runs the snow-survey program in Yosemite, has memorialized my charades of skier competence in his name for that crossing: Yochim's Plunge.

Fun times aside, the measurements we took on this trip were sobering, especially in hindsight. The snowpack we measured was on the light side, despite the foot of snow that fell on our second night. One dry year alone is not much cause for concern, but that one would become the beginning of the epic drought California was soon to suffer. Indeed, if we were to do the survey every year in late March (the one used for the April 1 winter summary) for the rest of the 2010s, we would have seen graphic evidence of that drought's severity, not only in the actual snowpack but also in the expedition itself. The year of no snow was most revealing, as one might expect. No snow, by definition, means that the surveyors had to walk most of the route, though there was still enough snow to ski the trip's final leg. That is the day on each trip that the team travels from the second of two patrol cabins they use (at Wilma Lake) to the highest site, Grace Meadow. That snow course ordinarily has at least eight feet of snow; in 2015, it had less than two. The Wilma Lake Cabin, which is almost completely buried by snow in a heavy winter, was naked, bereft of snow like the surrounding forest. With other recent winters also being warm enough to make skiing difficult or impossible for the other lower-elevation sites, Fincher now has skis stashed at the Wilma cabin for the one-day ski to Grace Meadow (the team walks the eighteen miles to the Wilma cabin, completing the four low- and mid-elevation snow courses, then skis to and from Grace Meadow and the Wilma cabin on the third day, and hikes the eighteen miles out on the fourth and final day). Warmer winters also prevent snow bridges from forming over creeks that were previously frozen over, forcing teams to wade them and, in one location, complete a dangerous traverse through avalanche terrain right above a cliff. Finally, even as early as 2012 (when I did the survey), the snowpack monitoring reflected a trend of declining SWE at lower elevations and increasing SWE at higher elevations; more recent winters have only accelerated the trend.[52]

Watching the snowpack numbers carefully are a variety of interests and entities, including the California Department of Water Resources, the state agency that manages California's water, in cooperation with other agencies like the NPS. No other state can match the size of California's agricultural economy, which produced close to $50 billion in crops in 2018. California farmers produce two-thirds of the nation's fruits and nuts, with over four hundred agricultural commodities statewide, almost all of which are irrigated. Many of the state's highest-value crops, including grapes, almonds, pistachios, and oranges (which together account for about 30 percent of the state's farm receipts in 2018) require a long-term availability of water.[53] Added to agriculture's water demand are those of nearly forty million people, the country's highest state population. To slake the combined thirst of residents and agriculture, various agencies have built America's longest and most elaborate system of impoundments, aqueducts, pipelines, tunnels, and pumping stations, transporting the water from where it falls (generally, in the northern half of the state) to where it's used (generally, in the southern half), and often over one or more mountain ranges.[54] All told, over 2,100 miles of major aqueduct line the California landscape; king among them is the 701-mile State Water Project, begun in 1960. It delivers water to twenty-five million residents from the Bay Area to Los Angeles, irrigates 750,000 acres, and has the single greatest water lift in the world, 1,926 feet up and over the Tehachapi Mountains to LA (which makes it the single greatest electricity consumer in the state).[55] Given this extensive reworking of the plumbing in Central and Southern California, it's perhaps not a surprise that few rivers south of Sacramento reach the sea, having been completely diverted for human uses. Collecting and storing much of the runoff on these rivers is an abundance of reservoirs—almost 1,500 statewide.[56] The state's largest reservoir has no dam, but is three times the size of the largest impoundment, the Sierra Nevada snowpack, which contains enough water in a typical year to supply all California residents.[57] Water is what has made all this possible, water from winter storms and the snow that falls in the Sierra—and water whose predictability and amount will likely be seriously altered the more we warm the climate.

Turning a final time to the Northwest Climate Toolbox for temperature and precipitation projections and using Snow Flat and Yosemite Valley for the representative locations, we see familiar patterns emerge from the model runs for the various seasons—patterns that, if anything, are more extreme than those in the other places we've examined.[58] Winter's average high temperatures at Snow Flat, already comfortably above freezing (41 degrees), will move well above it (51 degrees). Winter's overnight lows follow suit, moving from well below freezing (16 degrees) to more than a few nights that are above freezing (with an average of 25 degrees). Springtime will continue the trend away from snow country, with average lows moving from solidly frozen (23 degrees) to slushy, almost thawed (31 degrees), and high temperatures moving from cool (49 degrees) to springlike (59 degrees). Lest there be any doubt that winter will largely be a memory by the year 2100, autumn temperatures will also move away from Snowville, with lows moving from below freezing (30 degrees) to well above it (41 degrees). Daytime highs will virtually guarantee that any precipitation will be in liquid form, not frozen, moving from thinking about winter (55 degrees) to just dreaming about it (66 degrees). Summertime temperatures keep step with the other seasonal increases, moving from cool (67 degrees) to warm (78 degrees). Precipitation, meanwhile, increases modestly, but all in winter (twenty-eight to thirty-three inches), and partially at the expense of both fall (eight to ten inches) and spring (twelve to thirteen inches). Temperature projections for Yosemite Valley parallel these changes, but are 10–12 degrees warmer, except in winter, when the valley's tall cliffs cast long shadows, limiting solar influence and making the valley just 6 degrees warmer. Precipitation projections also parallel those of Snow Flat, just 20–25 percent less.[59] In short, Sierra Nevada (Spanish for "snowy range") will no longer be a fitting name for the range; Sierra Caliente or Sierra Seca ("warm mountains" and "dry mountains," respectively) might become more fitting.

And it will be both warm and dry—and wet too, though not at the same time. Using ten climate models to project the Sierra's future climate, a team of researchers found that the mountain range will lose 65–76 percent of its snowpack by century's end, an amount that exceeds by 40 percent the combined storage capacity of all eighty reservoirs on the Stanislaus, Tuolumne, Merced, and San Joaquin Rivers.[60] The scale of this loss is due to at least three factors, the first of which is the sizeable projected temperature increase. The second is that the Sierra's average winter high temperature, 41 degrees at Snow Flat, is already quite warm to both receive and hold on to snow. When a winter storm hits the Sierra, the volume and intensity of falling snow can depress temperatures, making it colder than it would be if snow were falling lightly.[61] This principle allows the relatively balmy Sierra Nevada to get considerably more snow than its mild temperatures would suggest. The third factor accounting for the significant loss is the high percentage of mid-elevation snow-gathering land in the Sierra land base: once global warming has moved the typical snowline (the elevation at which rain turns into snow) above nine thousand feet, the Sierra will lose most of its snow-covered terrain. Complicating the situation further will be a projected exaggeration of the high variability already present in California's precipitation from year to year. By the year 2100, extremely dry winters will be one and a half to two times as likely to occur, and extremely wet winters three times as likely, both compared to the historical frequency, as reported by a pair of researchers using thirty-four climate models.[62] We're likely seeing this already, with another team reporting a robust increase since 1949 in the strength and persistence of the northeastern Pacific high-pressure ridge that controls whether winter storms hit California. This increase, the team noted, did not come at the expense of wet winters.[63] Indeed, it comports with the recent reality in Yosemite, from the exceptional snow volume we were skiing across the Sierra on, to the unprecedented drought that began the next winter, and then to two more winters of exceptionally heavy snowfall. Such year-to-year variability, and the difficulties that brings to both human and natural systems, is what makes the Central and Southern California regions a climate change hot spot, an area where the magnitude of change is likely to be high, relative to other parts of the country.[64] The rules governing winter precipitation in Yosemite and California are being rewritten, thanks to us—but not necessarily in our favor.

Winter storms will still bring moisture to California and the Sierra, but more and more of it will be rain, not snow. In the arid West, moisture of any kind is usually welcome, but snow is generally preferred. Not only does the meltwater gradually soak into the soil (where rain can run off if it falls too hard), but in a state with a Mediterranean climate, the meltwater also essentially prolongs the wet season. Yosemite's forests enjoy a longer time period with abundant soil moisture, while California irrigators can leave water in the state's reservoirs till the runoff drops (depending on the snowpack, sometime in June or July). To prepare for the time when most of California's precipitation comes as rain, some have called for the state's leaders to build more reservoirs and thereby capture some of the snow-turned-rain. One site is available, and a few dams can be raised to capture more runoff, but otherwise, the best reservoir sites have been utilized.[65] The few remaining tend to have restrictions precluding development, like being in a national park or being a protected National Wild and Scenic River. Other methods to augment supply have similar problems. Groundwater pumping, already widespread in California, is exhausting supplies in some regions, with some wells in the San Joaquin Valley going down thousands of feet, and over five thousand new wells drilled during the recent drought. The current rate of drilling and pumping, which exceeds aquifer replenishment in most areas, is more akin to water mining than to sustainable use. There is increasing interest in groundwater banking, which is letting water soak into the ground during times of excess, and then withdrawing it in times of scarcity. Not all aquifers can be recharged in this manner, and it's uncertain where the water would come from, with all water already allocated.[66] Water conservation is also increasingly limited in its ability to help us cope with drought, for farmers have already widely adopted conservation equipment like drip irrigation, and cities have mandated low-flow faucets and toilets. These measures have allowed the state's population to grow by over 75 percent since the previous extreme drought (1976–1977), with almost no increase in overall municipal water demand.[67] Short-term conservation measures have increasingly relied on behavioral modifications, which are easily forgotten. Municipal water demand has "hardened," indicating it has little to give in times of scarcity. With no obvious silver bullet, then, the state's water future is not comforting, no matter the elaborate infrastructure. We appear to be on a collision course to a parched calamity, one that is entirely of our own making.

Returning to Yosemite, the forests we skied through are projected to see a 23 percent decrease in the soil moisture they need for optimal conditions for seedlings to establish themselves, across elevations, by the 2020–2049 timeframe, averaged through wet and dry years (the paper reporting these findings did not project beyond 2049). The range of moisture deficits will be 15–50 percent, and is expected to cause tree mortality, especially in trees already at their moisture limits, like mountain hemlock and western white pine.[68] The drier conditions and warmer temperatures will result in forests moving upslope, with oaks becoming the sole forest dominant below four thousand feet as conifers disappear there. In the mid-elevations (four thousand to seven thousand feet), oaks and ponderosa pines, the previous codominants in the lower elevations, will be the most successful in future "recruitment events" (times when moisture and temperatures are optimal for the seedlings of a given tree species to establish themselves enough to survive the next drought). Above seven thousand feet, Jeffrey and piñon pine, both generally found below that elevation now, will dominate recruitment events to come, while red fir, a stately tree with deep red bark often occurring in magnificent single-species stands in the snowbelt zones, will decline significantly. The other subalpine tree species, including lodgepole and whitebark pine, will see a modest decline in recruitment events, but will persist in the landscape.[69] For that matter, many of the mature trees present in the landscape will survive for decades and, for a few hardy individuals, even centuries, but they will no longer be able to perpetuate themselves in place, their seeds being cast into an environment that has become too warm and dry for undeveloped root systems. On a similar note, these projections are blind to the ability of tree species to migrate; some will need assistance to move successfully, but if that doesn't happen, the species could be left behind. In sum, the warmer and more erratic climate we're creating will cause forest changes on a dramatic scale, and at an accelerating pace, in a mountain range beloved to millions.

Catalyzing these changes in most areas will be fire. The declining snow cover is expected to result in more wildfires, both because forests will dry out earlier in the growing season and because snow on the ground suppresses lightning strikes by cooling the air above. By the 2020–2049 timeframe, the number of lightning-started fires is projected to increase by 19 percent, with the area of forest burned at high intensity increasing by 22 percent (again, the paper did not project beyond 2049).[70] Fire has not historically played as pivotal a role in the Sierra's high-elevation forests as it has in the northern Rockies forests, as suggested by the absence of serotinous cones in Yosemite's lodgepole pines. The Sierra Nevada does not get as many thunderstorms and associated lightning strikes as the Rockies, a factor that, in combination with the Sierra's more discontinuous fuels, relegates fire more to the ground than the canopy, which negates the value of serotinous cones. The dramatically different climate we are creating, however, will expand both the frequency and intensity of fire in High Sierra forests, along with smoke and blackened forests. The resulting landscape will be one that is new to the Sierra, one of black forests, seedling trees, senescing older trees, and other patches of forest confused about a rapidly changing climate. There will be pockets of this mountain range's once-grand forests, but by the end of the century, they will be increasingly fragmented, diseased, and hard to find. And as we'll see soon enough, the changes have already begun in earnest, and one doesn't need to look hard to find them.

Changes were happening inexorably to my breathing as well, which was slowly slipping away. I became more and more dependent on the BiPAP breathing assistance device, first increasing the time I used it in bed to the entire night plus my afternoon nap, then while writing in the mornings, and on and on. Then I began having trouble with it failing to respond when I inhaled; I would exhale but wouldn't find the comforting rush of air when I'd inhale. The respiratory technician knew that was the indication that my breathing had become too weakened for the BiPAP to suffice and that it was time to upgrade to a Trilogy ventilator, which is just a more sensitive version of the BiPAP. A Trilogy was indeed the solution, but it didn't stop the continuing decline in my breathing ability. However, the rate of my decline seemed to slow once I had finally resorted to using it twenty-four hours a day. Whether that low plateau is due to the Trilogy use or some other factor is anyone's guess, but the additional years I've enjoyed on planet Earth (time that continued use of the faltering BiPAP probably would not have given me) has allowed me to try to finish both this book in addition to my previous one, *Essential Yellowstone*, as well as spend more time with loved ones. The Trilogy also has a battery that lasts up to five hours, far exceeding that of the BiPAP, providing for activities outside the house, which has certainly improved my quality of life. I can still attend concerts of the St. Louis Symphony Orchestra, go for walks in nearby parks that have paved trails, and sit outside reading and relaxing in my parents' gardens. Quality-of-life improvements these certainly are for a terminally ill patient in the advanced stages of a debilitating disease.

As with many things medical, the improvements come at a cost. The most obvious is the face mask I have to wear to connect with the Trilogy, along with the air hose, which, together, make it look like I'm in intensive care, or like I'm an alien in a *Star Wars* movie. Another is my increasing dependence on this machine, both physical and psychological. Perhaps the main way we can tell that my breathing ability continues to weaken is how I respond when we remove the mask to brush my teeth or wash my face. When I first used the Trilogy, I had no problem without it for the few minutes it takes to do these tasks; two years later, I need refreshing breaths of the machine's reassuring pressured air every minute or so. I also need supplemental oxygen once in a while, usually in the evening when my energy is ebbing. There is also psychological dependence, prompted by the knowledge that I would probably not survive for more than a half hour without the machine's life-sustaining airflow. Without the Trilogy, I feel as though my chest is in a vice or, if I'm lying down, as though someone is sitting on my chest. Knowing that my breathing is totally dependent on a single machine has, quite understandably, given rise to a few anxiety or panic attacks, sometimes triggered by interruptions in the machine's flow, sometimes by the mere fear of interrupted airflow. Perhaps the greatest drawback to the Trilogy is needing it in the first place. This one machine, more than even my wheelchair, demonstrates to the world that my disability goes beyond the typical ailments that confine a person to such a chair. Not only am I restricted to pavement, but I cannot be away from an electric outlet for more than five hours, the life of the Trilogy battery. My travel options are increasingly limited to just the St. Louis area.

Indeed, my traveling days ended when I moved to Missouri a year after being diagnosed. In the first couple of years after I returned home, my parents and I enjoyed visiting many of our favorite attractions in eastern Missouri. All of them were within a two-hour drive of home, so we did them as day trips. These were some of my favorite childhood and adolescent haunts, places where we used to romp among elephantine granite boulders and swim in Ozarkian spring-fed streams. We didn't go beyond the two-hour distance limit, nor did we make any overnight trips. We did talk about some possible destinations, but dismissed them for reasons said (the wheelchair magnified road bumps and irregularities, turning some highways into bone-jarring, stomach-churning misadventures) and unsaid (bathroom issues). As my abilities declined—and as I grew dependent on breathing assistance—our outing range and time shrank, to the point that now, in my seventh year of living with ALS, we rarely travel even an hour from home, and winter's cold keeps me inside for days and sometimes weeks at a time. Given these constraints and the bulky medical equipment needed to care for me (a patient lift, shower chair, and hospital bed), it's clear that I can no longer travel. The amazing, uplifting landscapes of any park in this book are impossible to visit, their inspiration attainable to me only through recollections and through writing. "Dependence" seems too nice a descriptor for something as insidious as this prison sentence, one that seems more and more like solitary confinement.

Thinking back on my last day of the Trans-Sierra trip, if my group were to repeat the trek today, we would get a preview of the altered future of this terrain. We began that day at the Snow Flat Cabin in the snowy conifer forest, with so much snow on the ground that only the roof crown protruded above the white. It was to be the longest day of the Trans-Sierra trip, the first half skiing, the second half hiking down a long series of switchbacks to the floor of Yosemite Valley. The skiing was enjoyable, taking us through Yosemite's diverse mix of conifers and broad-leaved trees, most of them growing in open, magnificent stands that presented little difficulty to the skiers zipping through. We began in the stately red and white firs, schussed past huge Jeffrey and western white pines, switched back among hardy golden-cup and canyon live oaks on the sun-drenched slopes above Yosemite Valley, and finally finished in the towering sugar and ponderosa pines on the valley floor. We were completely on trails that day, just crossing the Tioga Road (where before we had used it for our route of travel). The skiing took us down 1,900 feet that morning; the hiking, another 2,700 feet that afternoon. The sizeable downhill was a pleasant challenge for everyone, and we took it as our abilities (the older skiers cautiously, the younger ones with more abandon) and equipment (namely me, with the damaged ski holding up) allowed. The last in our group made it to the trailhead at dusk, there to be greeted by a pizza dinner that the early-arriving skiers had picked up. It was a fantastic trip, one that offers still more insights into the changes we're causing through climate change.

For starters, we stopped for lunch at another cabin, the Snow Creek Cabin, built in 1929 by the Yosemite Park and Curry Company as part of an envisioned series of chalets modeled on those in the Alps. Shortly thereafter, the downhill-skiing operations at Yosemite's Badger Pass opened, taking the public interest rate away from Snow Creek. Just five years after building the two-story cabin, the company abandoned it. For the next thirty or forty years, the cabin was gradually forgotten, there being no trail to it to remind passersby of its existence (it's also in a dense forest, hidden from view from nearby mountaintops). According to an informal history posted in the cabin, some park employees rediscovered it in the 1970s, and it's been used for its original purpose ever since. Interest in alpine (downhill) skiing at Badger Pass has flagged with the opening of downhill-ski resorts featuring much more elevational drop than Badger, while interest in Nordic (cross-country) skiing has increased, explaining the cabin's resurgent popularity today. Meanwhile, Badger Pass continues to operate, one of only three downhill-ski facilities in the national park system (and by far the largest), though recent winters have brought too little snow for complete operating seasons, or, in the winter of no snow, an operating season at all (the first in its history). Nonetheless, the small resort soldiers on, its status as the West's oldest ski resort (tied with Sun Valley, Idaho; both opened their first ski lifts in 1936) providing momentum for its continued operation.[71] Badger Pass, along with the Snow Creek Cabin, are both testaments to the long history of our skiing pastimes in Yosemite. That these recreational traditions, both dependent on snow, are increasingly threatened by global warming is obvious. Indeed, the very buildings that bespeak of their presence in the Yosemite snowscape are at risk of destruction, especially the cabin, located as it is in a dense forest capable of sustaining a crown fire in the right conditions. It's not just natural and cultural resources at stake as we raise the global temperature—it's also long-established traditions like skiing and glacier research, and the meanings we find in them.

If we were to repeat this traverse across the Sierra Nevada today, we would see dramatic evidence of climate change near the top of the switchbacks down into Yosemite Valley. As the trail approaches the switchbacks, we would begin to see dead trees, dozens of them. Then, where the trail breaks out of forest and begins descending to the valley below, we would get our first view of the hundreds that died on the valley floor. The sudden change in solar exposure, from almost flat to nearly vertical on the south-facing side, makes the snow disappear (or appear, if you've just climbed up the switchbacks), along with the snow-loving conifers (replaced by sun-loving oaks). This is where the skier-turned-hiker pauses not just to remove skis and see dead trees but also to admire the colossi across the valley: Half Dome, rising 4,700 feet above the valley floor far below, and just east of it, Cloud's Rest, 1,000 feet higher still. Both granite monoliths show the handiwork of the glaciers once filling the Tenaya Creek and Merced River Valleys. The summit of Half Dome is rounded and flat, indicating that, as high as it is (8,844 feet above sea level), glaciers still covered it at some time in the past, rounding its mountaintop. With Half Dome and its neighbor Cloud's Rest—its north face scraped clean of any soil, manifesting such signs of a glacial past—juxtaposed with signs of our climate warming, the view from the top of the switchbacks is at once glacial and desiccating, cold and hot, icy and fiery.

As the epic drought unfolded across California, researchers began noticing substantial reductions in the canopy water content, which is the total amount of liquid water in the foliage of a tree, throughout forests statewide. One year into the drought, 283 million large trees (at least five inches in diameter at the four-foot level) exhibited a measurable amount of water loss, indicating they were drought stressed. A year later, the number had ballooned to 507 million, and another year saw the number swell to 565 million. This was the water year of no snow (water year 2015), so with two years of drought and heat already seen, the number of trees experiencing severe drought stress (a canopy water content loss of 30 percent or more) was 58 million. Even the usually snowy subalpine lodgepole pine and red fir forests showed canopy losses of 10 percent or more.[72] Losses of just 10 percent of canopy water can kill a tree, with figures exceeding 30 percent often fatal, so the ranks of dead trees grew by the tens of millions. Statewide, 11 million large trees had died by 2014, with that figure growing by 29 million in 2015. The year of no snow was especially hard on trees, with the casualty number more than doubling in 2016 to 62 million (102 million total to date), despite that water year's near-average snowpack (90 percent in Yosemite, and note that the statewide snowpack generally tracks with Yosemite's). Even the abundant snowfall the next year (173 percent of average in Yosemite) was not enough to overcome the cumulative stress of so many critically dry and warm years, with another 27 million trees perishing. The now-historic die-off continues, with 19 million trees dead in the below-average 2018 water year (though not markedly low, at 64 percent). The effects of that historic drought clearly continue to be felt in California's grand forests, with 148 million large trees dead by 2018, nearly a third of which succumbed after the drought itself came to an end.[73]

Dying trees are not the only way that drought's effects continue. The tree casualties were greatest on the west slope of the central and southern Sierra, primarily affecting trees at the lower end of their respective distributions, and in Yosemite, tapering off above six or seven thousand feet in elevation. With so much dry fuel on the landscape, it was probably inevitable that fires would start, and possibly become large. It was Yosemite and the national forest lands immediately west of it that saw some of the most significant fire activity, both during and after the drought. In 2013, the Rim Fire burned 257,000 acres on the northwest side of the park, including 77,000 acres within the park itself. This fire was the largest historically seen in the Sierra Nevada, and the third largest in California history at the time. Then, in 2018, the Ferguson Fire burned 97,000 acres on Yosemite's southwest side, including about 10,000 in the park itself. This fire resulted in the closure of Yosemite Valley for almost three weeks in the middle of the summer tourist season.[74] Almost all of the 350,000 acres incinerated in the two fires was below seven thousand feet in elevation, right in the forests most impacted by the epic drought. Clearly, that drought was far reaching and historic—and a window into the hot, fiery, and smoky future of Yosemite.

Thankfully, one tree that did not suffer many casualties in the recent drought was the Sierra's most famous tree, the giant sequoia. The big trees occur only in some seventy discrete groves on the Sierra's west slope, including two small groves and one large grove in Yosemite (the

Merced, Tuolumne, and Mariposa Groves, respectively). Our Trans-Sierra trip route did not approach any of Yosemite's groves, but the tale of climate change in the Sierra would be incomplete without discussing this iconic tree and its prospects in the face of climate change. However, rather than going to one of Yosemite's groves, let's take a brief side trip to Sequoia National Park and its Giant Forest, home to the largest tree on the planet, the General Sherman Tree. Giant Forest is gifted with four exquisite meadows, one of which I have long been enamored of. The forest of giants is also where I took my first ski tour after having had my first lesson at the nearby Wolverton ski area. Like the small ski lifts, Poma lifts, and rope tows that many western parks once had, Wolverton is now defunct, removed because it's no longer seen as appropriate in a national park, with dozens of much bigger resorts scarring the western landscape and leading to unfettered growth in the valleys nearby. (Yosemite is increasingly at odds with its contemporaries in continuing to operate Badger Pass, but given a few more years of business as usual, climate change will drive the last nail in its coffin of obsolescence and inappropriateness.) A final reason to divert southward is that I took one of my last hikes in Giant Forest, with my twin brother, just two months after receiving the diagnosis that would change my life forever. Let's join Jim and me on a fall afternoon, as we sat on a certain log in one of those meadows eating our lunch, as the drought was just beginning to take hold of the Golden State.

Of all the places I have been in California, by far the most powerful is that log in Giant Forest's Log Meadow. The log, near the head of the meadow, was once a sequoia enjoying a premier growing site, with the wet meadow providing more than enough water to slake the tree's five-hundred-gallon-a-day thirst.[75] But like most of the trees lining the edge of this and the other meadows, this tree ended up toppling into the meadow, the wet soil too soft to support a tree that measures age by the century (and sometimes millennia), not the mere decade. So the trees fall, producing convenient dining venues with their broad former trunks, six or eight feet in diameter (a skinny tree by sequoia standards; the Sherman Tree's thirty-six-foot base is more respectable).[76] I first found this particular log in 1990, when I was a seasonal park ranger stationed in Giant Forest. Once every week or two, I worked a shift in which I led both a morning and an afternoon nature walk at nearby Crescent Meadow. For lunch, I would hightail it over to my log in Log Meadow, shed my shirt and ranger hat (so other hikers wouldn't disturb me), and soak up the California sun for an hour. That wasn't all I imbibed, for the beauty and energy of that place were forces all their own.

As Jim and I found on that early November day, the colors I had once observed in the meadow's flowers had migrated to the round, shrubby, and yellow willows stationed here and there on the meadow edges, with one or two out in the meadow. Frost had turned the sedges and grasses of the meadow golden, but plenty of green could be seen above, in the foliage of the sequoias across the meadow. Green led our eyes skyward to the sequoia tree crown two hundred feet above, the tree's rounded canopy at odds with the pointed spires typical of coniferous trees. Down at our level, the exuberant sequoia foliage hid their red bark, along with a few large boulders, but here and there the colors of red and gray were unmasked. With the intense blue sky above complementing the rich color palette below, it was a kaleidoscope of color, solitude, and life.

If we were to sit on "my" log today, the forest across the meadow would appear little changed, despite the drought that was beginning when Jimmy and I had lunch there. In part, this is due to the site's inherent water availability, which enabled the nearby trees to survive four years of extreme aridity. Removed from its life-saving water, we would see a forest in trouble—except for the sequoias. While the other trees that occur with giant sequoias took big hits in the recent drought (white fir, 20–25 percent mortality; ponderosa pine, 50 percent; and sugar pine, a whopping 70 percent), less than 1 percent of the big trees died.[77]

The sequoia's distinctly superior performance during the drought is due to at least three factors. First, as mentioned above, the sequoia dominates the wetter locations. The same is true away from the meadows; in fact, abundant moisture is common to all sequoia groves (along with the Sierra's mild climate). The few that died were in areas of marginal soil water, such as lower elevations (where snowfall is less and evaporation greater), areas of low mature sequoia density (suggesting the site has inherent limitations), or steep hillsides (far from the nearest stream).

Sequoias like it wet, a strategy that serves them well when the inevitable drought strikes.[78]

Second, the sequoia employs a suite of drought-coping strategies. A sequoia's first line of defense is to close or restrict the stoma, which are the tiny openings that allow air into the needle's photosynthetic cells, usually found on the underside of the scale (sequoia foliage resembles the scales of juniper trees). Water vapor can also travel through the stoma, so by closing them, the tree reduces its water demand. In most dry years, stomatal closure is enough to get the tree through, which was the case for the first two years of this drought. But as 2014 brought unprecedented aridity and warmth, stomatal closure was no longer sufficient, and the trees risked drawing air into their xylem, which are the channels bringing water from the roots to the needles. Air in xylem is the kiss of death for a tree, as it destroys the hydrostatic connection between roots and foliage. To prevent this from happening, the sequoia drops its older needles, which are not as photosynthetically active as younger ones. Most evergreen conifers shed their needles after they've been on the tree a few years; sequoias simply accelerate the process in times of extreme drought (as do other conifers). Combined with stomatal closure, dropping older needles enabled almost all sequoias to weather the drought. That's the good news; the bad news is that these adjustments don't come without costs. Closing or restricting the stoma limits photosynthesis, and letting the older needles drop off certainly doesn't boost the tree's ability to capture solar energy. Consequently, droughts as severe as this one leave the big trees alive but weakened. Given a few years of more typical moisture, the trees can rebuild their photosynthetic capacity, but any xylem damage they suffered can be long term and difficult to repair.[79] Also, the scale of the foliage die-off in 2014 was without precedent in modern history (going back well over a century).[80] The sequoia's drought adaptations have helped it survive for millennia in a Mediterranean climate with markedly dry summers year after year, but we may well be pushing this noble survivor into extremes that even it cannot handle.

The same may also be true of insect pests, the absence of which is the third—and probably most significant—factor accounting for the sequoia's remarkable survivorship in California's recent drought. Until this drought, giant sequoias were only known to die by falling or by being scorched from hot updrafts from a fire. Cedar bark beetles had been known to kill a branch or two, or to invade a fallen branch, but the sequoia has otherwise been known for its amazing immunity to fatal insect attacks. In contrast, the proximate cause of most sugar pine and white fir deaths was bark beetles of one kind or another (with the drought being the ultimate cause). But during and after this drought, twenty to thirty sequoias died—standing—in Giant Forest, with no fire to blame. The evident cause of death, determined from beetle tunnels in branches fallen from the trees, was the cedar bark beetle. Although fatalities like this have probably happened in the past, this is the first documented case of bark beetles killing giant sequoias. Given severely stressed trees in unprecedented drought conditions, the beetles can apparently overcome this legendary survivor of a tree.[81] Notwithstanding its drought-coping strategies, the implications for the sequoia in the hotter future we're creating are not sanguine. Indeed, the future for the sequoias has much in common with that facing McClure Glacier, the pika and other small mammals and birds, and a host of other flora and fauna: rising temperatures, increasing weather extremes, shifting biomes, more frequent fire, and possible extinctions. The sequoia as a species will probably persist in some, perhaps many, of its seventy groves, but it won't be unscathed, as recently demonstrated in Yosemite's southern counterpart.

Not long after I was diagnosed with ALS, I read an article in the local paper about a fellow with the same disease and his fundraising efforts in support of the ALS Association, the primary nonprofit advocacy organization supporting families with an ALS sufferer. The reporter who wrote the article was a good friend of mine, and knowing my situation, he had asked the guy, Tom Santi, if he would welcome my call and possible visit. Tom said yes, and just a few days later, I had both called and visited him. Misery loves company as the saying goes, so Tom and I readily bonded, despite (as I would eventually find) our divergent political beliefs and the fact that he was a climate change denier. Mutual respect and our common affliction saw us through these differences, even after I moved out of California and we became email buddies. Five years passed with hardly a month of silence between us, sometimes not even a day. Coping with our common affliction was the most frequent topic of discussion for us, but there were many others, particularly his wife and three grown children, and my writing. I shared draft manuscripts with him (which may be the reason he eventually renounced his denier beliefs), while he told me that he estimated that as many as two hundred hummingbirds were finding sustenance at the numerous feeders he and his family had on their property. We found we both liked to cook and, like many ALS victims, had both been physically fit when diagnosed (he was a long-distance runner, and I, a long-distance hiker). More seriously, two or three times he told me he thought his time on Earth was drawing to a close, but each time, he rallied, drawing strength from the love of others. It was a friendship born of mutual suffering and loss, both of us prisoners in our own bodies.

And then in August 2019, as I was writing this chapter—about the park on his doorstep—he didn't rally. He had written me a few weeks earlier, again telling me that he thought his time might be near. Something about his message seemed different than his previous ones, so I replied differently, congratulating him on finally making it to the end of his sentence, if indeed he did. My email then moved on to other things, without telling him that I valued our friendship and hoped he didn't leave us yet (which were all things I'd told him previously). Tom didn't reply, and then there passed an unusually long silence between us, long enough to make me think I should touch base with him. Trying to make headway on this long chapter, though, I deferred writing him. Eventually, he beat me to it, or so I thought. When I opened the email, I saw that it was actually his wife, Donna, writing to tell me that he'd died in his sleep a few nights earlier. Oh, how I didn't want that news, despite—or because of—my last email to him. Tom had had the disease a year or two longer than I, so we both knew that he'd probably get released before I did, but that does not temper the rudeness of the surprise when you get the bad news. I have wept several times for my friend and email buddy, and regret not writing him when I was suspecting a change was afoot in him. May he rest in peace and, wherever he is, find the long-distance running shoes and trails of his best dreams.

Tom is the third and last person I have met who have escaped their ALS-ruined bodies (besides Anne Rayburn, I met one other acquaintance of my parents). I know of others who suffer its indignities, but have not met them in person. In that sense, I am now alone, though I have many healthy friends and family members who keep me good company. Perhaps more than ever, I am resigned to living out my days in semi-solitary, partly because it's very hard to imagine a wonder drug or treatment that could both halt the disease progression and repair the massive neurological damage ALS has caused in my body. I spend my days trying to make the best of my situation, writing about the places I love and the most urgent existential threat to them in history. Doing this helps bring me back to these amazing and beautiful places, to relive the solitude and stillness of an April morning sitting atop the Tioga Pass cabin, to watch the wild waters of Yellowstone Lake calm as the setting sun sinks into it, and to feel the exuberance of life in Log Meadow amid the most powerful assemblage of the world's largest tree. If there are any places that energize the body, mind, and soul, they are these places, at least for me. If there are any places that calm the hurried, that soothe the wounded, that quiet the troubled, they are these islands of tranquility. And if there are any places that offer hope for the terminally ill—and hope that nature can prosper if we just give it a chance—they are places like these, even if we can only visit them in our memories. National parks don't have a monopoly on such places; Tom, for example, enjoyed watching the hummingbirds zooming around his house. The parks do, however, have some of the most powerful and moving landscapes in the world, such as the places featured in this book. They are an American birthright, one of the most profound gifts to future generations that any society can give.

And they have still more to give to those who visit, and visit deeply. The parks, at the most basic level, are extraordinary venues to contemplate the meanings enshrined within them, the values of nature, and our place in worlds both natural and human. They are landscapes of learning, places that beg us to ponder the immense forces behind their creation, places that have stories to tell of how they continue to be molded by myriad forces, places that showcase our continuously evolving relationship to nature.[82] They are places of inspiration that move artists to paint canvases so glorious they stir a nation to protect spectacular wild landscapes for all time, that inspire musicians to compose pieces so evocative that the changing moods of nature practically douse us with an afternoon thunderstorm, and that motivate scientists to spend lifetimes researching natural secrets so profound that no amount of study can plumb their full depths. Finally, they are places that call us to look beyond our own interests and consider the benefit of others present and future, places of altruism that make us think about the greater good, places that demand we take action against threats to our exceptional birthright.

And they are screaming: *Take action against climate change, NOW, before we forever lose their wonders, both tangible and intangible.*

CONCLUSION

Hiking down from Blue Glacier in Olympic National Park, I wondered what the viewpoints would look like a decade or two later, given the warmth that had me in shorts and a T-shirt at just midmorning. How far would the glacier recede? Will the people who make the trek to that place know the difference between what they see and what was? What about those who make the trip thirty or forty years from now, when the glacier is significantly smaller? At what point—if ever—will the view from the upper overlook no longer inspire us? The mountain won't disappear, but will it be as jaw dropping without Blue Glacier and its impressive ice field? And if I, an employee of the very agency charged with protecting the parks, hadn't known about the glacier beforehand, how many future hikers will recognize the difference between the naked mountain they see and the formerly ice-clad mountain fastness? In other words, if some beauty will survive the melt-off, why worry about global warming, especially since many of us (perhaps most) won't know the difference?

Such questions are difficult to answer effectively, in part because we don't know what the future will bring. But I think we know some of the answers: beauty reduced is beauty diminished, for without its glaciers, the mountains in Olympic will no longer be unique and will resemble dozens of other mountain ranges in the West. Beauty reduced also violates the code we have as caretakers of the earth, for it means we're not passing the planet down to our children in better shape than we received it from our parents. It violates the golden rule, too—to do unto others as we would have them do unto us—for it means we are damaging communal properties. It disrespects our fellow creatures, because glaciers provide year-round cold water to the streams and rivers that sustain an abundance of fish and birds and mammals and amphibians. Finally, it suggests an arrogance that is at odds with the attitude of humility and restraint one should have in the presence of anything powerful and complex, a heedless disregard for the consequences of our actions, known and unknown, direct and indirect, short term and long term. There are other reasons, some of which have been highlighted in this book, but these are the most important in my mind. Besides being a view out of place, the view of Blue Glacier was also a scene out of time, knowing that the glacier's years are numbered. It was like looking at the last of an endangered species, thrilling to see the creature in the flesh, but sad to know that even if we directed all our personal efforts into saving the species, the animal would probably still go extinct. The view into the heart of Olympic, then, was simultaneously thrilling and disquieting.

As the planet warms more, some plants and animals may come to find their habitat too hot or dry, forcing them to migrate to find suitable conditions for their growth. If that is impossible for them, they may die out, possibly everywhere. Our crops may fail, victims of the same changes. Melting glaciers and ice caps in Greenland and Antarctica will raise ocean levels and flood many of the world's most populous cities; already, the Atlantic Ocean on the Eastern Seaboard has risen five or six inches, producing recurring minor flooding during especially high tides. Also, warming the climate energizes oceans, producing more powerful and destructive storms, like the 2017 hurricanes in the Gulf of Mexico. The list of possible effects, present and predicted, is long indeed; a full treatment of them is beyond the scope of this book, which focuses on the national parks.

And that picture is disturbing. In addition to the western parks that I've discussed in earlier chapters, all of the nation's parks are facing challenges from climate change. For example, the Everglades face inundation from unchecked sea-level rise. The forests of the Great Smoky Mountains are increasingly at risk of larger and more frequent forest fires. In the Alaskan parks, as the permafrost of the tundra melts, frozen carbon escapes into the atmosphere via several different pathways. Scientists are just beginning to understand those mechanisms, so they have not been included in the climate change models. Estimates are that only 5–15 percent of the frozen carbon will escape to the atmosphere, which may not sound like much. But when you multiply that number by the vast amount of tundra in Alaska, Canada, and Siberia, it's equivalent to the total amount of carbon we have in the atmosphere. This is quite worrisome.

Yet, today, the society that was responsible for setting aside these wonderful places for future generations is turning its back on the future. The human contribution to climate change is still a matter of debate among the Republican Party and others who don't believe scientific findings that conflict with their way of seeing the world. They might as well debate whether smoking causes lung disease or whether the earth is round, for the research proving the human role in warming the planet is just as solid and conclusive. In forcing a debate where there is nothing debatable, and thereby delaying action to stave off further warming, Republicans like Donald Trump have led the way toward a climate never before seen in the human experience, and at a rate that has also never been seen. The perils for humanity are many, serious, and destabilizing, including excessive heat, crippling drought, food and water insecurity, warfare over these effects, coastal flooding and refugee displacement, and many other harrowing problems. It is the height of irresponsibility to deflect attention from this most urgent existential threat.

What can be done to reduce the impact of climate change? If we get serious about reducing our emissions now, impacts could be about half as bad as those discussed in this book. Keep in mind that the business-as-usual numbers I presented were the average values, not the range of them. Some people are taking action. Many nations have aggressive climate change policies. Even without consistent national leadership in the United States, dozens of states and hundreds of cities have endorsed the goals of the Paris Agreement. In the national parks, managers are pursuing efforts in accordance with the Climate Smart Conservation program. Many companies and individuals are also taking steps to reduce their carbon footprint. But, without responsible leadership from national policy makers, these efforts will not be enough.

My own experience with ALS enhances my sensitivity to these potential changes. If we don't curtail our carbon emissions soon—and radically curb them—my experience may well become the *universal* experience.

As the glaciers melt, forests burn, and wildlife move or die out across the West, all in the face of climate change, our collective memories of the unimpaired national parks will gradually fade, much like my increasingly distant recollections. And much like my inability to travel to the parks, we'll be increasingly limited in when we can visit parks, due to the growing prevalence of fire, smoke, and other hazards. This projection of the future visitor experience, if you will, also extends to the landscape we will see; just as I look out on the altered landscape of Missouri, future visitors to the parks in this book will see landscapes that have been significantly changed. However, just as there are some pockets of naturally functioning lands in Missouri, the parks will hold on to patches of naturally functioning forests and other vegetation types. "Naturally functioning" may be illusory, with no landscape completely immune to global warming, but some of these pockets will at least display nature's forces equal to, or exceeding, the hand of humanity, both contemporary and prehistoric. In this and other ways, the parks will probably always inspire, but it will probably become increasingly difficult to find inspiration in them, with their resource impairments more and more obvious to all. In sum, climate change is to the body politic as ALS is to those who suffer its effects: a growing threat to the parks and to the inspiration we draw from them.

ACKNOWLEDGMENTS *William R. Lowry*

Mike Yochim never quit. He was determined to finish this book and literally died trying, sitting at his desk, typing. But Mike also knew that he might not be able to complete it and so asked me, in the event of his death, to do so. Mike and I did not discuss acknowledgments before he passed, but I think I have a pretty good idea of what he would say.

First, Mike would want to acknowledge his family, especially his parents, Jim and Jeanne. They provided incredible and constant support for him during the time he was writing the manuscript, always with patience and love. Similarly, his brothers and their families, as well as his friends and the ALS Association, were caring and supportive. We express much gratitude to all the wonderful people who helped Mike on the difficult road he traveled in dealing with ALS. Mike would be proud that proceeds from the sales of this book will go to the ALS Association.

Second, I'm confident that Mike would want to acknowledge the men and women of the National Park Service, who strive to protect the remarkable places described in this book. Many of them provided insight and information to Mike while he was working on the manuscript. Just to mention one here, we are grateful to Jim Donovan of Yosemite National Park, who helped with the photos. Others are mentioned specifically in the manuscript. Mike cared deeply about the parks and valued the support and friendship of others in the NPS, who do as well.

Third, Mike would want to thank those of us who helped finish this book after he passed away in February 2020. I coordinated that effort and appreciated a great deal of help. Mike's brother Brian Yochim provided very helpful assistance on several aspects of the process. Brian would like to thank Jill and Ellis Yochim for their support while he worked to select and organize the photos for this book and perform edits. Mike's friend Eric Compas generously donated his skill and expertise to making the maps. Stephanie Croghan helped with the endnotes. Mike's parents were involved throughout. I personally want to thank my wife, Lynn, and the good people at the University of New Mexico Press, especially Michael Millman and Marie Landau. We also want to thank the outside readers, Jack Loeffler and Todd Wilkinson, for their strong and insightful endorsement of the manuscript. Their enthusiasm gave us all encouragement to finish the book that Mike started. Finally, Mike's family graciously insisted on thanking me for my efforts. I'll simply say that I am proud to have been Mike's friend and a part of this book.

NOTES

Foreword

1. Yochim and Lowry, "Creating Conditions for Policy Change."

Introduction

1. Cornell Lab of Ornithology, "All about Birds: Canyon Wren."
2. Water-data.com, Lake Powell Water Database.
3. US Global Change Research Program, Third National Climate Assessment.
4. Ault et al., "Assessing the Risk of Persistent Drought."
5. The Grand Staircase–Escalante National Monument, gutted by the Trump administration in 2017, is immediately adjacent to Glen Canyon National Recreation Area.
6. Gonzalez et al., "Disproportionate Magnitude of Climate Change"; US Geological Survey, "Retreat of Glaciers"; National Park Service (NPS), North Cascades National Park, "Climate Change Resource Brief."
7. American Cancer Society, "Lifetime Risk of Developing or Dying From Cancer"; American Heart Association and American Stroke Association, "Heart Disease and Stroke Statistics."
8. Sax, *Mountains without Handrails*; Runte, *Yosemite*.

Chapter One

1. McNulty, *Olympic National Park*; NPS, Olympic National Park, "Temperate Rain Forests."
2. McNulty, *Olympic National Park*; NPS, Olympic National Park, "Animals."
3. McNulty, *Olympic National Park*.
4. Riedel et al., "Glacier Status and Contribution to Streamflow"; William Baccus, email message to author, December 20, 2018; NPS, Olympic National Park, "Glaciers and Climate Change."
5. McNulty, *Olympic National Park*.

6. Roe, Baker, and Herla, "Centennial Glacier Retreat."

7. National Aeronautics and Space Administration (NASA), "Global Climate Change."

8. Marlier et al., "2015 Drought in Washington State"; Abatzoglou, Rupp, and Mote, "Seasonal Climate Variability and Change"; Monahan and Fisichelli, "Recent Climate Change Exposure of Olympic National Park"; NPS, "Vital Signs Overviews."

9. Rasmussen and Conway, "Estimating South Cascade Glacier (Washington, USA) Mass Balance."

10. Baccus, *Park Interpreter's Guide to the Climate of Hurricane Ridge*; Baccus, Larrabee, and Lofgren, *North Coast and Cascades Network Climate Monitoring Report*; William Baccus, email message to author, December 20, 2018. See also NPS, "Vital Signs Overview."

11. Marlier et al., "2015 Drought in Washington State."

12. Abatzoglou, Rupp, and Mote, "Seasonal Climate Variability and Change."

13. East et al., "Channel-Planform Evolution."

14. Ibid.

15. Luce and Holden, "Declining Annual Streamflow Distributions."

16. van Mantgem et al., "Widespread Increase of Tree Mortality."

17. Abatzoglou, Rupp, and Mote, "Seasonal Climate Variability and Change"; Mote et al., "Perspectives on the Causes of Exceptionally Low 2015 Snowpack."

18. Woodward, Schreiner, and Silsbee, "Climate, Geography, and Tree Establishment."

19. C. Ray et al., "Recent Stability of Resident and Migratory Landbird Populations."

20. Intergovernmental Panel on Climate Change, *Climate Change 2014*; NASA, "Climate Change" and "Paleoclimatology"; Wikipedia, "Climate Change," https://en.wikipedia.org/wiki/Climate_change, and "Ice Core," https://en.wikipedia.org/wiki/Ice_core, both accessed April 28, 2018.

21. Yochim, "Aboriginal Overkill Overstated"; Yochim, *Yellowstone and the Snowmobile*.

22. Powell, "Consensus on Anthropogenic Global Warming"; Wikipedia, "Scientific Consensus on Climate Change," accessed May 15, 2018, https://en.wikipedia.org/wiki/Scientific_consensus_on_climate_change?wprov=srpw1_0. For examples of other such papers, see Oreskes, "Beyond the Ivory Tower"; Cook et al., "Quantifying the Consensus on Anthropogenic Global Warming."

23. NASA, "Global Climate Change"; Wikipedia, "Scientific Consensus on Climate Change."

24. Reidel et al., "Glacier Status and Contribution to Streamflow."

25. East et al., "Channel-Planform Evolution"; Mantua, Tohver, and Hamlet,

"Climate Change Impacts on Streamflow Extremes"; former Yosemite hydrologist Jim Roche, email message to author, November 29, 2019.

26. Griggs, "Sea-Level Rise for the Coasts."

27. Lindsey, "Climate Change."

28. NPS, Olympic National Park, "Fire History."

29. Burwell, "Environmental Factors Influencing Fire Behavior."

30. Oregon Historical Society, "Oregon Encyclopedia: Tillamook Burn"; Wikipedia, "Tillamook Burn," accessed May 3, 2018, https://en.wikipedia.org/wiki/Tillamook_Burn; Perry, "How Bad Can a Fire Get in Oregon?"

31. For this discussion, I generally relied on the University Corporation for Atmospheric Research (UCAR) Center for Science Education, "Climate Modeling"; NOAA Climate.gov, "Climate Models"; and Hegewisch and Abatzoglou, "Future Time Series."

32. UCAR, Climate Modeling"; NOAA Climate.gov, "Climate Models"; Hegewisch and Abatzoglou, "Future Time Series."

33. Hegewisch and Abatzoglou, "Future Time Series."

34. Rupp and Li, "Less Warming Projected during Heavy Winter Precipitation."

35. William Baccus, email message to author, December 18, 2019.

36. Hegewisch and Abatzoglou, "Future Time Series."

37. Ibid.

38. East et al., "Channel-Planform Evolution."

39. Thoma, Munson, and Witwicki, "Landscape Pivot Points and Responses."

40. Gergel et al., "Effects of Climate Change on Snowpack."

41. Elsner et al., "Implications of 21st Century Climate Change."

42. Littell et al., "Forest Ecosystems, Disturbance, and Climatic Change."

43. Sheehan, Bachelet, and Ferschweiler, "Projected Major Fire and Vegetation Changes."

44. Westerling, "Increasing Western US Forest Wildfire Activity."

45. Olympic National Park fire management officer Todd Rankin, email message to author, February 1, 2019.

46. Littell et al., "Forest Ecosystems, Disturbance, and Climatic Change."

47. Dennison et al., "Large Wildfire Trends in the Western United States"; Westerling, "Increasing Western US Forest Wildfire Activity."

48. Metcalfe, "Watch the Pacific Northwest Change."

49. Nakawatase and Peterson, "Spatial Variability in Forest Growth"; Peterson, Peterson, and Ettl, "Growth Responses of Subalpine Fir"; McNulty, *Olympic National Park*.

Chapter Two

1. Of the 277 river miles, up to 50 are submerged in Lake Mead.

2. NPS, Grand Canyon National Park, "Park Statistics," "Reptiles," and "Mammals."

3. NPS, Grand Canyon National Park, "Geologic Formations"; Wikipedia, "Geology of the Grand Canyon Area," accessed June 3, 2018, https://en.wikipedia.org/wiki/Geology_of_the_Grand_Canyon_area.

4. Wikipedia, "Dendrochronology," https://en.wikipedia.org/wiki/Dendrochrnology, and "Dendroclimatology," https://en.wikipedia.org/wiki/Dendroclimatology, both accessed June 17, 2018.

5. Salzer and Kipfmueller, "Reconstructed Temperature and Precipitation"; NPS, "Climate Change in the Southwest."

6. Salzer and Kipfmueller, "Reconstructed Temperature and Precipitation."

7. Udall and Overpeck, "Twenty-First Century Colorado River"; McCabe et al., "Evidence That Recent Warming Is Reducing Upper Colorado."

8. Udall and Overpeck, "Twenty-First Century Colorado River."

9. Vano et al., "Understanding Uncertainties"; Udall and Overpeck, "Twenty-First Century Colorado River."

10. Gonzalez et al., "Southwest."

11. Wikipedia, "Amyotrophic Lateral Sclerosis," accessed January 8, 2020, https://en.wikipedia.org/wiki/Amyotrophic_lateral_sclerosis.

12. Grand Canyon Trust, "Stopping Grand Canyon Escalade"; Sottile, "Navajo Nation Votes Down Controversial Hotel."

13. Wikipedia, "Marble Canyon Dam," https://en.wikipedia.org/wiki/Marble_Canyon_Dam, and "Bridge Canyon Dam," https://en.wikipedia.org/wiki/Bridge_Canyon_Dam, both accessed July 2, 2018.

14. Powell, *Dead Pool*, 109–11; Anderson, *Polishing the Jewel*, 66–67; Wikipedia, "David Brower," accessed July 6, 2018, https://en.wikipedia.org/wiki/David_Brower; Sierra Club, "David Brower (1912–2000)."

15. Anderson, *Polishing the Jewel*, 66–67.

16. NPS, "Theodore Roosevelt Quotes," accessed May 19, 2021, https://www.nps.gov/thro/learn/historyculture/theodore-roosevelt-quotes.htm.

17. Webb and Magirl, "Changing Rapids of Grand Canyon."

18. Easterling et al., "Precipitation Change in the United States."

19. Sources vary on the amount of warming seen in the Grand Canyon, depending in part on the size of the area analyzed, with the park and its immediate surroundings warming more than the region as a whole. The first two of the following analyses report an increase of about 4 degrees, using only the park's records and nearby weather stations, while the rest of them report an increase of about

half that for the region as a whole. Analyses include Monahan and Fisichelli, "Recent Climate Change Exposure of Grand Canyon National Park"; Nicholas Fisichelli, "Climate Change Trends for Planning (backcountry management) at Grand Canyon National Park," white paper for NPS use, 2013, attachment to an October 30, 2018, email message to author from Grand Canyon employee Jean Palumbo; Jepson et al., "Evaluation of Temperature and Precipitation Data"; Lenart, *Global Warming in the Southwest*; Dvey, Redmond, and Simeral, *Weather and Climate Inventory*.

20. Monahan and Fisichelli, "Climate Exposure of US National Parks."

21. Thoma, Norris, and Lauck, "Satellite-Based Vegetation Condition and Phenology"; NPS, "Climate Change in the Southwest."

22. NPS, "Climate Change in the Sonoran Desert."

23. Williams et al., "Forest Responses to Increasing Aridity and Warmth"; Dennison et al., "Large Wildfire Trends."

24. Shaw, Steed, and DeBalder, "Forest Inventory and Snalysis (FIA) Annual Inventory."

25. NPS, "Climate Change on the Southern Colorado Plateau"; Floyd, Romme, and Hanna, "Fire History and Vegetation Pattern in Mesa Verde"; Breshears et al., "Regional Vegetation Die-Off"; Dvey, Redmond, and Simeral, *Weather and Climate Inventory*.

26. Andrews et al., *Describing Past and Future Soil Moisture*; Fisichelli, "Climate Change Trends for Planning."

27. Redmond, Forcella, and Barger, "Declines in Pinyon Pine Cone Production."

28. Adams et al., "Temperature Response Surfaces for Mortality Risk."

29. NPS, "Climate Change in the Southwest"; Cook, Ault, and Smerdon, "Unprecedented 21st-Century Drought Risk" (source of quote); Dvey, Redmond, and Simeral, *Weather and Climate Inventory*; Fisichelli, "Climate Change Trends for Planning."

30. Romme et al., "Historical and Modern Disturbance Regimes."

31. Colorado Natural Heritage Program, "Pinyon-Juniper"; NPS, "Birds and Climate Change."

32. Flatley and Fulé, "Are Historical Fire Regimes Compatible?"

33. NPS, "Climate Change in the Southwest"; Rehfeldt et al., "Empirical Analyses of Plant-Climate Relationships."

34. Diffenbaugh et al., "Fine-Scale Processes."

35. NPS, "Plants: Grand Canyon Species List."

36. Weiss and Overpeck, "Is the Sonoran Desert Losing Its Cool?"; Barrows and Murphy-Mariscal, "Modeling Impacts of Climate Change."

37. Patrick Gonzalez, "Anthropogenic Climate Change in Joshua Tree National

Park, California," NPS white paper dated July 26, 2019, provided by email message to author from Ronda Newton, January 7, 2020.

38. McAuliffe and Hamerlynck, "Perennial Plant Mortality."

39. Hereford, Webb, and Longpre, "Precipitation History and Ecosystem Response."

40. Iknavan and Beissinger, "Collapse of a Desert Bird Community."

41. Riddell et al., "Cooling Requirements."

42. Albright et al., "Mapping Evaporative Water Loss."

43. Cruz-Mcdonnell and Wolf, "Rapid Warming and Drought."

44. Brooks and Matchett, "Spatial and Temporal Patterns of Wildfires."

45. Syphard, Keeley, and Abatzoglou, "Trends and Drivers of Fire Activity."

46. Abatzoglou and Kolden, "Climate Change in Western Deserts."

47. Sweet et al., "Congruence between Future Distribution Models and Empirical Data"; Gonzalez, "Anthropogenic Climate Change."

48. Cook et al., "Unprecedented 21st-Century Drought Risk."

49. Schmidt, *Grand Canyon*; Wikipedia, "Uinkaret Volcanic Field," accessed August 6, 2018, https://en.wikipedia.org/wiki/Uinkaret_volcanic_field; NPS, Grand Canyon National Park, "Geologic Activity."

50. Union of Concerned Scientists, "Lake Mead"; Patterson, "Should Iconic Lake Powell Be Drained?"; Davis, "Risks to Lake Mead, Colorado River." A full discussion of the complicated history and management of the Colorado River is outside the scope of this book; see James Powell, *Dead Pool*, 2008.

Chapter Three

1. NPS, Glacier National Park, "Nature & Science."

2. The name Pete Whalen is a pseudonym.

3. NPS, Glacier National Park, "Plants."

4. NPS, Glacier National Park, "Geology."

5. Ibid.; Wikipedia, "Stromatolite," accessed September 8, 2018, https://en.wikipedia.org/wiki/Stromatolite.

6. US Geological Survey (USGS), "Repeat Photography Project"; Fagre et al., "Glacier Margin Time Series."

7. Hall and Fagre, "Modeled Climate-Induced Glacier Change"; Brown, Harper, and Humphrey, "Cirque Glacier Sensitivity,"; USGS, "Retreat of Glaciers"; Fagre et al., "Glacier Margin Time Series."

8. O'Neel et al., "Reanalysis of the US Geological Survey Benchmark Glaciers."

9. Pederson et al., "Century of Climate and Ecosystem Change."

10. Hall and Fagre, "Modeled Climate-Induced Glacier Change"; Pederson et al., "Climatic Controls."

11. Pederson et al., "Century of Climate and Ecosystem Change."

12. Pederson et al., "Climatic Controls."

13. Monahan and Fisichelli, "Recent Climate Change Exposure of Glacier National Park."

14. Pederson et al., "Climatic Controls."

15. Ibid.

16. USGS, "Retreat of Glaciers."

17. USGS, "Glacier Monitoring Studies."

18. Pederson et al., " Climatic Controls"; Pederson et al., "Unusual Nature of Recent Snowpack Declines."

19. Pederson, Betancourt, and McCabe, "Regional Patterns and Proximal Causes."

20. Marzeion et al., "Attribution of Global Glacier Mass Loss."

21. Pederson et al., "Century of Climate and Ecosystem Change."

22. Pederson et al., "Climatic Controls."

23. Hegewisch and Abatzglou, "Future Time Series."

24. Pederson et al., "Climatic Controls."

25. Larson et al., "Modelling Climate Change Impacts."

26. NPS, Glacier National Park, "Ice Patch Archeology Resource Brief."

27. Fagre et al., "Glacier Margin Time Series."

28. Roush, Munroe, Fagre, "Development of a Spatial Analysis Method"; USGS, "Repeat Photography Project" (see the five views under the "Vegetation Change" subheading); NPS, Glacier National Park, "Climate Change."

29. Clark, Harper, and Fagre, "Glacier-Derived August Runoff."

30. Muhlfeld et al., "Climate Change Links Fate of Glaciers"; Giersch et al., "Climate-Induced Glacier and Snow Loss."

31. D'Angelo and Muhlfeld, "Factors Influencing the Distribution of Native Bull Trout."

32. Glacier National Park Conservancy, "Black Swifts."

33. Hansen et al., "Influence of Streamflow on Reproductive Success."

34. Jones, Muhlfeld, and Marshall, "Projected Warming Portends Seasonal Shifts of Stream Temperatures."

35. Lesica, "Arctic-Alpine Plants Decline over Two Decades"; NPS, Glacier National Park, "Plants."

36. Bloom et al., "Compounding Consequences of Wildfire and Climate Change."

37. Moyer-Horner et al., "Predictors of Current and Longer-Term Patterns"; NPS, Glacier National Park, "Pikas Resource Brief" (note that this brief provides a much lower population estimate based on Moyer-Horner's work, without explanation for the difference from his published estimate, so I went with the published one); Garrett et al., "Pikas in Peril."

38. Glacier Park Foundation, "Glacier Park Fires of 2003" and "Fires of 2003."

39. NPS, "Wildfire History at Glacier National Park"; NPS, Glacier National Park, "Fire History."

40. Hegewisch and Abatzglou, "Future Time Series."

41. Egan, *Big Burn*; Forest History Society, "1910 Fires"; Wikipedia, "Great Fire of 1910," accessed October 3, 2018, https://en.wikipedia.org/wiki/Great_Fire_of_1910.

42. Egan, *Big Burn*.

43. NPS, Glacier National Park, "National Park Service Releases Review of Fire," "Howe Ridge Fire Update," and "Fire-Fueled Finds."

44. Earthwatch Institute, "Climate Change, Huckleberries, and Grizzly Bears."

Chapter Four

1. The impulse to preserve nationally special landscapes dates back to at least 1783, when Mongolia set aside Bogd Khan Mountain (Wikipedia, "Bogd Khan Mountain," accessed February 21, 2020, https://en.wikipedia.org/wiki/Bogd_Khan_Mountain). In the United States, the preservation movement's first notable success came in 1864, when the federal government deeded Yosemite Valley and the nearby Mariposa Grove of giant sequoias to the State of California; these areas were eventually incorporated into Yosemite National Park. In 1872, Yellowstone National Park was established, but since Wyoming, Montana, and Idaho were all territories, the federal government retained ownership and gave the park the National Park title that has since inspired the preservation movement worldwide.

2. Most of this paragraph draws from Yellowstone's website, in particular the following pages, all of which can be found under NPS, Yellowstone National Park: "Plants," "Nature and Science," "Mammals," "Geology," "Research," and "Yellowstone Lake." For the record-low temperatures, see Frank H. Anderson to the Superintendent, memorandum, January 31, 1949, loose within box N-158, Yellowstone National Park Archives, Gardiner, Montana; Wikipedia, "Rogers Pass," accessed February 14, 2019, https://en.wikipedia.org/wiki/Rogers_Pass.

3. According to social psychologist Milton Rokeach's 1973 *The Nature of Human Values*.

4. Yochim, *Week in Yellowstone's Thorofare*.

5. Yochim, *Essential Yellowstone*.

6. Haines, *Yellowstone Story*, 1:35–38; Wikipedia, "John Colter," accessed February 21, 2019, https://en.wikipedia.org/wiki/John_Colter.

7. NPS, Yellowstone National Park, "1988 Fires."

8. I detail most of these reports in my second book *Protecting Yellowstone: Science and the Politics of National Park*.

9. Romme, "Fire and Landscape Diversity in Subalpine Forests" (quote); Romme and Despain, "Historical Perspective on the Yellowstone Fires"; Turner, Romme, and Tinker, "Surprises and Lessons from the 1988 Yellowstone Fires."

10. Romme et al., "Twenty Years after the 1988 Yellowstone Fires"; NPS, Yellowstone National Park, "1988 Fires."

11. This paragraph is based primarily on Romme et al, "Twenty Years after the 1988 Yellowstone Fires," and Romme et al., "Deterministic and Stochastic Processes." For additional detail, see Wallace, ed., *After the Fires*; Romme et al., "Establishment, Persistence, and Growth of Aspen"; Romme et al., "Rare Episode of Sexual Reproduction in Aspen"; Romme et al., "Aspen, Elk, and Fire in Northern Yellowstone"; Schoennagel, Turner, and Romme, "Influence of Fire Interval and Serotiny."

12. Westerling, "Increasing Western US Forest Wildfire Activity"; Dennison et al., "Large Wildfire Trends."

13. I was unable to find an authoritative source providing a list of largest fires by state. For that reason, I consulted Wikipedia, "List of Wildfires" (accessed March 4–6, 2019, https://en.wikipedia.org/wiki/List_of_wildfires) and the National Interagency Fire Center, "Wildfires Larger Than 100,000 Acres (1997–2018)" (accessed March 4–6, 2019, https://www.nifc.gov/fireInfo/fireInfo_statistics.html). Note that Oregon may have seen a much larger fire in 1845, when a fire burned as much as 1.5 million acres, but I was unable to find a reliable source for this. Also, for the following three states, I found Wikipedia to be deficient, so I used the National Interagency Fire Center and another internet source, as follows. Idaho: PBS, "America's Most Devastating Wildfires" (https://www.pbs.org/wgbh/americanexperience/features/burn-worst-fires/); Nevada: Daniel Rothberg, "'It's Gone, It's Gone': Nation's Largest Wildfire in Nevada Devastates Ranches, Sage Grouse" (July 12, 2018, https://thenevadaindependent.com/article/its-gone-its-gone-nations-largest-wildfir-in-nevada-devastates-ranches-sage-grouse); Wyoming: Rodeo Rick, "How Big Were Wyoming's Worst Wildfires?" (August 1, 2016, https://mycountry955.com/how-big-were-wyomings-worst-wildfires/), all accessed March 5, 2019.

14. Dennison et al., "Large Wildfire Trends."

15. Mote et al., "Dramatic Declines in Snowpack"; Wikipedia, "List of Largest Reservoirs in the United States," accessed March 10, 2019, https://en.wikipedia.org/wiki/List_of_largest_reservoirs_in_the_United_States.

16. Pederson, Betancourt, and McCabe, "Regional Patterns and Proximal Causes."

17. Westerling, "Increasing Western US Forest Wildfire Activity."

18. Pederson et al., "Regional Patterns and Proximal Causes."

19. Abatzoglou and Williams, "Impact of Anthropogenic Climate Change."

20. Westerling et al., "Warming and Earlier Spring."

21. Monahan and Fisichelli, "Recent Climate Change Exposure of Yellowstone National Park."

22. Chang and Hansen, "Historic and Projected Climate Change."

23. Tercek, Rodman, and Thoma, "Trends in Yellowstone's Snowpack."

24. Ibid.

25. Romme et al.. "Twenty Years after the 1988 Yellowstone Fires"; Stevens-Rumann et al., "Evidence for Declining Forest Resilience"; Dennison et al., "Large Wildfire Trends"; Westerling, "Increasing Western US Forest Wildfire Activity."

26. There are several elevations on various websites for Eagle Peak; this one appears most common.

27. Smedes and Prostka, *Stratigraphic Framework of the Absaroka Volcanic Supergroup*; Hiza, "Geologic History of the Absaroka Volcanic Province."

28. Neither Smedes and Prostka, *Stratigraphic Framework*, nor Hiza, "Geologic History of the Absaroka Volcanic Province," had an answer to this question, nor did I find anything in a half dozen internet searches.

29. Though we might have been able to see Lake Yellowstone Hotel, far to the northwest, it's not discernible in any of my photos.

30. MacDonald, *Before Yellowstone*; Nabokov and Loendorf, *Restoring a Presence*.

31. Teton Mountains: NPS, Grand Teton National Park, "Glacier Monitoring"; Wind River Mountains: Cheesbrough et al., "Estimated Wind River Range (Wyoming, USA) Glacier Melt"; Beartooth Mountains: Seifert et al., "Monitoring Alpine Climate Change."

32. See Portland State University, "Glaciers of Wyoming," for some context.

33. Hansen et al., "Shifting Ecological Filters."

34. Donato, Harvey, and Turner, "Regeneration of Lower-Montane Forests."

35. Thoma et al., "Patterns of Primary Production & Ecological Drought."

36. Diaz and Eischeid, "Disappearing 'Alpine Tundra' Köppen Climatic Type."

37. Wells, "Clark's Nutcracker and Whitebark Pine"; Chang, Hansen, and Piekielek, "Patterns and Variability of Projected Bioclimatic Habitat"; Thoma, Shanahan, and Irvine, "Climatic Correlates of White Pine Blister Rust."

38. Buotte et al., "Climate Influences on Whitebark Pine Mortality."

39. Raffa, Powell, and Townsend, "Temperature-Driven Range Expansion of an Irruptive Insect"; Shanahan et al., "Whitebark Pine Mortality"; and Buotte et al., "Climate Influences on Whitebark Pine Mortality."

40. Shanahan et al., "Whitebark Pine Mortality."

41. Macfarlane, Logan, and Kern, "Innovative Aerial Assessment."

42. Chang et al., "Patterns and Variability of Projected Bioclimatic Habitat."

43. Shanahan et al., "Whitebark Pine Mortality"; Thoma, Shanahan, and Irvine, "Climatic Correlates of White Pine Blister Rust."

44. Keane et al., *Restoring Whitebark Pine Ecosystems.*

45. Buotte et al., "Climate Influences on Whitebark Pine Mortality."

46. McKinney, Fiedler, and Tomback, "Invasive Pathogen Threatens Bird-Pine Mutualism."

47. Whitlock and Hostetler, "Past Warm Periods."

48. Hegewisch and Abatzoglou, "Future Time Series." April 1 snowpack date is from Chang and Hansen, "Historic and Projected Climate Change."

49. Tercek and Rodman, "Forecasts of 21st Century Snowpack"; Yochim, *Yellowstone and the Snowmobile.*

50. Yellowstone supervisory vegetation management specialist Roy Renkin, email message to author, March 15, 2019.

51. Westerling et al., "Continued Warming Could Transform Greater Yellowstone"; Turner, "Climate Change and Fire."

52. Tercek, "Nowcasting & Forecasting Fire Severity."

53. Millspaugh, Whitlock, and Bartlein, "Variations in Fire Frequency and Climate."

54. Gross et al., "Historical and Projected Climates."

55. Cataldo and Smith, "2016 Yellowstone Fires."

56. Turner et al., "Short-Interval Severe Fire."

57. Nelson et al., "Landscape Variation in Tree Regeneration"; Nelson et al., "Simulated fire Behaviour."

58. Turner et al., "Short-Interval Severe Fire."

59. Yellowstone supervisory vegetation management specialist Roy Renkin, PowerPoint presentation ("Fire Behavior and Climate Change") attached to email message to author, March 15, 2019.

60. Stevens-Rumann et al., "Evidence for Declining Forest Resilience."

61. Harvey, Donato, and Turner, "High and Dry"; Harvey, Donato, and Turner, "Burn Me Twice."

62. Hansen and Turner, "Origins of Abrupt Change?"

63. Hansen et al., "It Takes a Few to Tango."

64. Turner et al., "Climate Change and Novel Disturbance Regimes."

65. Cochran and Berntsen, "Tolerance of Lodgepole and Ponderosa Pine Seedlings."

66. Hansen and Phillips, "Potential Impacts of Climate Change"; Rehfeldt et al., "Empirical Analyses of Plant-Climate Relationships"; Whitlock and Hostetler, "Past Warm Periods." See also Hansen et al., "Changing Climate Suitability for Forests."

67. Halofsky et al., " Nature of the Beast."

68. NPS, Yellowstone National Park, "Amphibians."

69. Ray et al., "Influence of Climate Drivers"; Ray et al., "Monitoring Greater Yellowstone."

70. Chang and Hansen, "Historic and Projected Climate Change." Note that the authors were unclear about whether the 1.4–3.2 degrees of stream warming projected for midcentury is in addition to the 1.8 degrees already observed in the Yellowstone area. Given the significant warming projected for the region, I assumed it would be.

71. NPS, Yellowstone National Park, "Yellowstone Cutthroat Trout," and "Lake Trout"; USGS, "FAQ on Invasive Lake Trout."

72. Mills et al., "Camouflage Mismatch"; NPS, Yellowstone National Park, "Canada Lynx."

73. Peacock, "Projected 21st Century Climate Change"; NPS, Yellowstone National Park, "Canada Lynx" and "Wolverine."

74. Moose sightings are from Thorofare Ranger Station logbooks; the seventy-six sightings number is from Eugene Young to Mother, letter, October 23, 1938, T. Eugene Young Papers, msc. 129, box 1, both at Yellowstone National Park Archives, Gardiner, Montana. Moose population estimates are from NPS, Yellowstone National Park, "Moose."

75. Becker, "Habitat Selection, Condition, and Survival"; Henningsen et al., "Distribution and Prevalence of *Elaeophora schneideri*"; Cusick, "Rapid Climate Changes"; NPS, Yellowstone National Park, "Moose." Moose in the Yellowstone ecosystem are also being weakened by an increase in brain worms. In Minnesota, the increased contact between whitetail deer (which are increasingly overwintering in areas that were once the winter province of moose alone) and moose explains the worm's growing prevalence in moose, though that connection has not yet been established in the Yellowstone area.

76. I have used pseudonyms for her and husband.

Chapter Five

1. NPS, Yosemite National Park, "Natural Resource Statistics."

2. Yosemite compliance specialist Lisa Acree, email message to author, July 21, 2019.

3. Wikipedia, "*Pinus lambertiana*," accessed July 20, 2019, https://en.wikipedia.org/wiki/Pinus_lambertiana; American Forests, "Champion Trees National Register," accessed July 20, 2019, https://www.americanforests.org/get-involved/americas-biggest-trees/champion-trees-national-register.

4. NPS, Yosemite National Park, "Natural Resource Statistics"; NPS, Yellowstone National Park, "Plants."

5. Wahrhaftig et al., "Extent of the Last Glacial Maximum (Tioga) Glaciation"; Glazner and Stock, *Geology Underfoot in Yosemite National Park.*

6. Yosemite wilderness specialist Mark Fincher, email message to author, July 15, 2019.

7. New Mexico state historic preservation officer (and former Tioga Pass entrance supervisor) Jeff Pappas, email message to author, August 8, 2019.

8. Nesmith et al., "Whitebark and Foxtail Pine."

9. Former Yellowstone and Grand Teton National Parks soundscape technician Shan Burson, email message to author, August 8, 2019.

10. Formichella et al., "Yosemite National Park Acoustic Monitoring."

11. NPS, "Parsons Memorial Lodge"; Wahrhaftig et al., "Extent of the Last Glacial Maximum (Tioga) Glaciation."

12. Righter, *Battle Over Hetch Hetchy*; NPS, "Parsons Memorial Lodge;" Wikipedia, "David Brower," accessed August 29, 2019, https://en.wikipedia.org/wiki/David_Brower.

13. Monahan and Fisichelli, "Recent Climate Change Exposure of Yosemite National Park."

14. Grinnell and Storer, *Animal Life in the Yosemite*; Wikipedia, "Belding's Ground Squirrel," accessed August 31, 2019, https://en.wikipedia.org/wiki/Belding%27s_ground_squirrel.

15. Morelli et al., "Anthropogenic Refugia Ameliorate."

16. Stewart et al., "Revisiting the Past."

17. C. Moritz et al., "Impact of a Century of Climate Change."

18. Rowe et al., "Spatially Heterogeneous Impact of Climate Change."

19. Tingley et al., "Push and Pull of Climate Change."

20. Forister et al., "Compounded Effects of Climate Change."

21. Lutz, van Wagtendonk, and Franklin, "Climatic Water Deficit."

22. McIntyre, "Twentieth-Century Shifts"; van Mantgem et al., "Widespread Increase of Tree Mortality."

23. Lutz et al., "Climatic Water Deficit."

24. Schwartz et al., "Increasing Elevation of Fire."

25. Bunn, Graumlich, and Urban, "Trends in Twentieth-Century Tree Growth."

26. Millar et al., "Response of Subalpine Conifers."

27. Wahrhaftig et al., "Extent of the Last Glacial Maximum (Tioga) Glaciation."

28. Stock, "Vanishing Ice, Vanishing History."

29. Ibid.; Basagic and Fountain, "Quantifying 20th Century Glacier Change."

30. Wikipedia, "Old Man of the Mountain," accessed September 15, 2019, https://en.wikipedia.org/wiki/Old_Man_of_the_Mountain.

31. Wikipedia, "Medlicott Dome," https://en.wikipedia.org/wiki/Medlicott_Dome, "Pywiack Dome," https://en.wikipedia.org/wiki/Pywiack_Dome, both accessed September 20, 2019.

32. Glazner and Stock, *Geology Underfoot in Yosemite*; Stine, "Extreme and Persistent Drought."

33. Graumlich, "1000-Year Record of Temperature and Precipitation"; Bunn et al., "Trends in Twentieth-Century Tree Growth."

34. All snowpack figures for Yosemite are from Yosemite wilderness specialist Mark Fincher, email message to author, July 15, 2019. To simplify this discussion, I averaged the figures he provided for the Tuolumne and Merced River drainages, which each drain about half the park.

35. Swain et al., "Trends in Atmospheric Patterns"; California Department of Water Resources, "Current Conditions."

36. Berg and Hall, "Anthropogenic Warming Impacts."

37. Diffenbaugh, Swain, and Touma, "Anthropogenic Warming."

38. Griffin and Anchukaitis, "How Unusual Is the 2012–2014 California Drought?"

39. Berg and Hall, "Anthropogenic Warming Impacts"; Belmecheri et al., "Multi-century Evaluation"; and Diffenbaugh, Swain, and Touma, "Anthropogenic Warming."

40. Mote et al., "Perspectives on the Causes of Exceptionally Low 2015 Snowpack."

41. Robeson, "Revisiting the Recent California Drought."

42. Mote et al. "Perspectives on the Causes of Exceptionally Low 2015 Snowpack."

43. Berg and Hall, "Anthropogenic Warming Impacts."

44. Seager et al., "Causes of the 2011 to 2014 California Drought."

45. Mote et al. "Perspectives on the Causes of Exceptionally Low 2015 Snowpack."

46. Swain et al., "Trends in Atmospheric Patterns."

47. Shukla et al., "Temperature Impacts on the Water Year 2014 Drought."

48. Williams, "Contribution of Anthropogenic Warming."

49. Ibid.

50. "Avalanche Death Delays Pass Opening," *Los Angeles Times*, June 15, 1995, https://www.latimes.com/archives/la-xpm-1995-06-15-mn-13537-story.html.

51. Andrews, *Hydrology of the Sierra Nevada*.

52. Ibid.; Mark Fincher, email messages to author, July 18, 2019, September 30, 2019, and October 3, 2019.

53. California Department of Food and Agriculture, "California Agricultural Production Statistics."

54. California Department of Water Resources, "State Water Project"; Wikipedia, "Water in California," accessed October 9, 2019, https://en.wikipedia.org/wiki/Water_in_California. See also Reisner, *Cadillac Desert*.

55. Wikipedia, "State Water Project," https://en.wikipedia.org/wiki/California_State_Water_Project, "Hetch Hetchy," https://en.wikipedia.org/wiki/Hetch

_Hetchy, "Los Angeles Aqueduct," https://en.wikipedia.org/wiki/Los_Angeles_Aqueduct, and "Mokelumne Aqueduct," https://en.wikipedia.org/wiki/Mokelumne_Aqueduct (all accessed October 19, 2019); American Society of Civil Engineers, "Colorado River Aqueduct"; US Bureau of Reclamation Mid-Pacific Region, "Central Valley Project."

56. Wikipedia, "List of Dams and Reservoirs in California," accessed October 18, 2019; Public Policy Institute of California, "Dams in California."

57. Natural Resources Defense Council, "California Snowpack and the Drought."

58. I first used Tuolumne Meadows for the representative upper elevation, but I found the results skewed by its cold air sink. Cold air is heavier than warm air, so at night, cold air moves down-valley, pooling where topography interrupts its continued movement. In this case, the cold air from both Lyell Canyon (just below the former glacier) and the Tioga Pass area pool in the meadows, making it colder than its elevation would otherwise be. While this attribute confers some resilience to the forests in that vicinity, it erodes the meadows' value as a representative location for the Yosemite high country. For that reason, I used Snow Flat, which does not have this confounding attribute.

59. Hegewisch and Abatzoglou, "Future Time Series."

60. Gergel et al., "Effects of Climate Change on Snowpack."

61. Rupp and Li, "Less Warming Projected during Heavy Winter Precipitation."

62. Berg and Hall, "Increased Interannual Precipitation Extremes."

63. Swain et al., "Trends in Atmospheric Patterns."

64. Diffenbaugh, Giorgi, and J.Pal, "Climate Change Hotspots."

65. Skelton, "California Should Stop Thinking about More Dams."

66. Bloch, "California Must Abandon 535,000 Acres of Prized Farmland"; Seager et al., "Causes of the 2011 to 2014 California Drought."

67. Diffenbaugh, Swain, and Touma, "Anthropogenic Warming."

68. Lutz, van Wagtendonk, and Franklin, "Climatic Water Deficit."

69. Liang, Hurteau, and Westerling, "Response of Sierra Nevada Forests." Whitebark and foxtail projections are from Moore et al., "Climate Change and Tree-Line Ecosystems."

70. Lutz et al., "Climate, Lightning Ignitions, and Fire Severity."

71. NPS, Yosemite National Park, "Snow Creek Cabin"; Wikipedia, "Badger Pass Ski Area," https://en.wikipedia.org/wiki/Badger_Pass_Ski_Area, "Sun Valley, Idaho," https://en.wikipedia.org/wiki/Sun_Valley,_Idaho, both accessed November 3, 2019.

72. Asner et al., "Progressive Forest Canopy Water Loss."

73. USFS, "Survey Finds 18 Million Trees Died."

74. Wikipedia, "Rim Fire," https://en.wikipedia.org/wiki/Rim_Fire, "Ferguson Fire," https://en.wikipedia.org/wiki/Ferguson_Fire, both accessed November 7, 2019; Mariposa County, "Post Ferguson Fire."

75. Ambrose et al., "Leaf- and Crown-Level Adjustments."

76. NPS, Sequoia and Kings Canyon National Parks, "General Sherman Tree."

77. Stephenson et al., "Which Trees Die during Drought?"

78. Stephenson et al., "Patterns and Correlates of Giant Sequoia Foliage Dieback"; Paz-Kagan et al., "Landscape-Scale Variation in Canopy Water Content."

79. Ambrose et al., "Leaf- and Crown-Level Adjustments."

80. Martin et al., "Remote Analysis of Canopy Water Content."

81. Nate Stephenson, email messages to author, June 7, 2019, and November 14, 2019.

82. For this idea I am indebted to Tom Vale, my doctoral advisor, who with his wife, Geraldine, wrote an entire manuscript on the topic, but chose not to publish it.

BIBLIOGRAPHY

Archives and Depositories

Yellowstone National Park Archives, Gardiner, Montana.

Books, Articles, and Reports

Abatzoglou, J. T., and C. A. Kolden. "Climate Change in Western Deserts: Potential for Increased Wildlife and Invasive Annual Grasses." *Rangeland Ecology and Management* 64, no. 5 (2011): 471–78.

Abatzoglou, J. T., D. E. Rupp, P. W. Mote. "Seasonal Climate Variability and Change in the Pacific Northwest of the United States." *Journal of Climate* 27, no. 5 (2014): 2125–42.

Abatzoglou, J. T., and A. P. Williams. "Impact of Anthropogenic Climate Change on Wildfire across Western US Forests." *Proceedings of the National Academy of Sciences USA* 113 (2016): 11770–75, 2016.

Adams, H. D., G. A. Barron-Gafford, R. L. Minor, A. A. Gardea, L. P. Bentley, D. J. Law, D. D. Breshears, N. G. McDowell, and T. E. Huxman. "Temperature Response Surfaces for Mortality Risk of Tree Species with Future Drought." *Environmental Research Letters* 11 (2017): 115014.

Albright, Thomas P., Denis Mutiibwa, Alexander. R. Gerson, Eric Krabbe Smith, William A. Talbot, Jacqueline J. O'Neill, Andrew E. McKechnie, and Blair O. Wolf. "Mapping Evaporative Water Loss in Desert Passerines Reveals an Expanding Threat of Lethal Dehydration." *Proceedings of the National Academy of Sciences* 114, no. 9 (February 2017): 2283–88.

Ambrose, A., W. Baxter, R. E. Martin, G. P. Asner, E. Francis, K. R. Nydick, and T. Dawson. "Leaf- and Crown-Level Adjustments Help Giant Sequoias Maintain Whole-Tree Hydraulic Integrity during Severe Drought." *Forest Ecology and Management* 419–20 (2018): 257–67.

Anderson, Michael F. *Polishing the Jewel: An Administrative History of Grand Canyon National Park.* Grand Canyon, AZ: Grand Canyon Association, 2000.

Andrews, C., J. Bradford, J. Norris, J. Gremer, M. Duniway, S. Munson, L. Thomas, and M. Swan. *Describing Past and Future Soil Moisture in the Pinyon-Juniper Woodland Community in Grand Canyon National Park*. Fort Collins, CO: NPS, 2018. Accessed December 20, 2018. https://irma.nps.gov/DataStore/DownloadFile/601910.

Andrews, Edmund D. *Hydrology of the Sierra Nevada Network National Parks: Status and Trends*. Natural Resource Report NPS/SIEN/NRR—2012/500. Fort Collins, CO: NPS, 2012.

Asner, Gregory P., Phillip G. Brodrick, Christopher B. Anderson, Nicholas Vaughn, David E. Knapp, and Roberta E. Martin. "Progressive Forest Canopy Water Loss during the 2012–2015 California Drought." *Proceedings of the National Academy of Sciences* 113 (2016): E249–55.

Ault, T. R., J. E. Cole, J. T. Overpeck, G. T. Pederson, and D. M. Meko. "Assessing the Risk of Persistent Drought Using Climate Model Simulations and Paleoclimate Data." *Journal of Climate* 27, no. 20 (2014): 7529–49.

Baccus, William D. *A Park Interpreter's Guide to the Climate of Hurricane Ridge, Olympic National Park: Climate Summary for Water Years 2000 to 2017*. Natural Resource Report NPS/NCCN/NRR—2018/1714. Fort Collins, CO: NPS, 2018. Accessed November 20, 2018. http://npshistory.com/publications/olym/nrr-2018-1714.pdf.

Baccus, W. D., M. Larrabee, and R. Lofgren. *North Coast and Cascades Network Climate Monitoring Report: Olympic National Park; Water Year 2012*. Natural Resource Data Series NPS/NCCN/NRDS—2016/1002. Fort Collins, CO: NPS, 2016. Accessed December 29, 2018. https://irma.nps.gov/DataStore/DownloadFile/545568.

Barrows, Cameron W., and Michelle L. Murphy-Mariscal. "Modeling Impacts of Climate Change on Joshua Trees at Their Southern Boundary: How Scale Impacts Predictions." *Biological Conservation* 152 (2012): 29–36.

Basagic, H.J., and A. G. Fountain. "Quantifying 20th Century Glacier Change in the Sierra Nevada, California." *Arctic, Antarctic, and Alpine Research* 43 (2011): 317–30.

Becker, Scott A. "Habitat Selection, Condition, and Survival of Shiras Moose in Northwest Wyoming." Master's thesis, University of Wyoming, 2008.

Belmecheri, S., F. Babst, E. R. Wahl, D. W. Stahle and V. Touret. "Multi-century Evaluation of Sierra Nevada Snowpack." *Nature Climate Change* 6, no. 1 (2016): 2–3.

Berg N., and A. Hall. "Anthropogenic Warming Impacts on California Snowpack during Drought." *Geophysical Research Letters* 44 (2017): 2511–18.

———. "Increased Interannual Precipitation Extremes over California under Climate Change." *Journal of Climate* 28 (2015): 6324–34.

Bloom, Trevor D. S., Aquila Flower, Michael Medler, and Eric G. DeChaine.

"The Compounding Consequences of Wildfire and Climate Change for a High-Elevation Wildflower (Saxifraga austromontana)." *Journal of Biogeography* 45, no. 12 (December 2018): 2755–65.

Breshears, D. D., N. S. Cobb, P. M. Rich, K. P. Prices, C. D. Allen, R. G. Balice, W. H. Romme, J. H. Kastens, M. L. Floyd, J. Belnap, J. J. Anderson, O. B. Myers, and C. W. Meyer. "Regional Vegetation Die-Off in Response to Global-Change-Type Drought." *Proceedings of the National Academy of Science* 102, no. 42 (2005): 15144–48.

Brooks, M. L., and J. R. Matchett. "Spatial and Temporal Patterns of Wildfires in the Mojave Desert, 1980–2004." *Journal of Arid Environments* 67 (2006): 148–64.

Brown, J., J. Harper, and N. Humphrey. "Cirque Glacier Sensitivity to 21st Century Warming: Sperry Glacier, Rocky Mountains, USA." *Global Planet Change* 74 (2010): 91–98.

Bunn, Andrew G., Lisa J. Graumlich, and Dean L. Urban. "Trends in Twentieth-Century Tree Growth at High Elevations in the Sierra Nevada and White Mountains, USA." *Holocene* 15, no. 4 (2005): 481–88.

Buotte, Polly C., Jeffrey A. Hicke, Haiganoush K. Preisler, John T. Abatzoglou, Kenneth F. Raffa, and Jesse A. Logan. "Climate Influences on Whitebark Pine Mortality from Mountain Pine Beetle in the Greater Yellowstone Ecosystem." *Ecological Applications* 26, no. 8 (2016): 2507–24.

Chang, Tony, and Andrew Hansen. "Historic and Projected Climate Change in the Greater Yellowstone Ecosystem." *Yellowstone Science* 23, no. 1 (2015): 14–19.

Chang, T., A. J. Hansen, and N. Piekielek. "Patterns and Variability of Projected Bioclimatic Habitat for *Pinus albicaulis* in the Greater Yellowstone Area." *PLoS ONE* 9, no. 11 (2014): e111669.

Cheesbrough, Kyle, Jake Edmunds, Glenn Tootle, Greg Kerr, and Larry Pochop. "Estimated Wind River Range (Wyoming, USA) Glacier Melt Water Contributions to Agriculture." *Remote Sensing* 1 (2009): 818–28.

Clark, Adam M., Joel T. Harper, and Daniel B. Fagre. "Glacier-Derived August Runoff in Northwest Montana." *Arctic, Antarctic, and Alpine Research* 47, no. 1 (2015): 1–16.

Cochran, P. H., and Carl M. Berntsen. "Tolerance of Lodgepole and Ponderosa Pine Seedlings to Low Night Temperatures." *Forest Science*, 19, no. 4 (December 1973): 272–80.

Colorado Natural Heritage Program. "Pinyon-Juniper: Impacts and Adaptation Strategies in a Changing Climate." Colorado State University: Fort Collins, 2018.

Cook, B. I., T. R. Ault, and J. E. Smerdon. "Unprecedented 21st-Century Drought Risk in the American Southwest and Central Plains." *Science Advances* 1, no. 1 (2015): e1400082.

Cook, John, Dana Nuccitelli, Sarah A. Green, Mark Richardson, Bärbel Winkler, Rob Painting, Robert Way, and Andrew Skuce. "Quantifying the Consensus on Anthropogenic Global Warming in the Scientific Literature." *Environmental Research Letters* 8, no. 2 (January 2013): 024024.

Cruz-Mcdonnell, K. K., and B. O. Wolf. "Rapid Warming and Drought Negatively Impact Population Size and Reproductive Dynamics of an Avian Predator in the Arid Southwest." *Global Change Biology* 22, no. 1 (2016): 237–53.

D'Angelo, Vincent S., Clint C. Muhlfeld, and Clint C. Muhlfeld. "Factors Influencing the Distribution of Native Bull Trout and Westslope Cutthroat Trout in Streams of Western Glacier National Park, Montana." *Northwest Science* 87, no. 1 (January 2013). https://doi.org/10.3955/046.087.0101.

Dennison, P. E., S. C. Brewer, J. D. Arnold, and M. A. Moritz. "Large Wildfire Trends in the Western United States, 1984–2011." *Geophysical Research Letters* 41 (2014): 2928–33.

Diaz, H. F., and J. K. Eischeid. "Disappearing 'Alpine Tundra' Köppen Climatic Type in the Western United States." *Geophysical Research Letters* 34 (2007): L18707.

Diffenbaugh, N. S., F. Giorgi, and J. S. Pal. "Climate Change Hotspots in the United States." *Geophysical Research Letters* 35 (2008): L16709.

Diffenbaugh, N. S., J. S. Pal, R. J. Trapp, F. Giorgi, and S. H. Schneider. "Fine-Scale Processes Regulate the Response of Extreme Events to Global Climate Change." *Proceedings of the National Academy of the United States of America* 102, no. 44 (2005): 15774–78.

Diffenbaugh, Noah S., Daniel L. Swain, and Danielle Touma. "Anthropogenic Warming Has Increased Drought Risk in California." *Proceedings of the National Academy of Sciences* 112, no. 13 (March 2015): 3931–36.

Donato, D. C., B. J. Harvey, and M. G. Turner. "Regeneration of Lower-Montane Forests a Quarter-Century after the 1988 Yellowstone Fires: A Fire-Catalyzed Shift in Lower Treelines?" *Ecosphere* 7, no. 8 (2016): e01410–01426.

Dvey, C. A., K. T. Redmond, and D. B. Simeral. *Weather and Climate Inventory, National Park Service, Southern Colorado Plateau Network*. Natural Resource Technical Report NPS/SCPN/NRTR—2006/007. Fort Collins, CO: NPS, 2006.

East, Amy E., Kurt J. Jenkins, Patricia J. Happe, Jennifer A. Bountry, Timothy J. Beechie, Mark C. Mastin, Joel B. Sankey, and Timothy J. Randle. "Channel-Planform Evolution in Four Rivers of Olympic National Park, Washington, USA: The Roles of Physical Drivers and Trophic Cascades." *Earth Surface Processes and Landforms* 42 , no. 7 (June 2017): 1009–1152.

Easterling, D. R., J. R. Arnold, T. Knutson, K. E. Kunkel, A. N. LeGrande, L. R. Leung, R. S. Vose, D. E. Waliser, and M. F. Wehner. "Precipitation Change in the United States." In *Climate Science Special Report: Fourth National Climate Assessment*, Vol. I, edited by D. J. Wuebbles, D. W. Fahey, K. A. Hibbard, D. J. Dokken, B. C. Stewart, and T. K. Maycock, 207–30. Washington, DC: US Global Change Research Program, 2017.

Elsner, M. M., L. Cuo, N. Voisin, Jeffrey S. Deems, Alan F. Hamlet, Julie A. Vano, Kristian E. B. Mickelson, Se-Yeun Lee, and Dennis P. Lettenmaier. "Implications of 21st Century Climate Change for the Hydrology of Washington State." *Climatic Change* 102 (2010): 225–60.

Fagre, D. B., L. A. McKeon, L. A. Dick, and A. G. Fountain. "Glacier Margin Time Series (1966, 1998, 2005, 2015) of the Named Glaciers of Glacier National Park." USGS data release, 2017. Accessed September 2018. https://doi.org/10.5066/F7P26WB1

Flatley, W. T., and P. Z. Fulé. "Are Historical Fire Regimes Compatible with Future Climate? Implications for Forest Restoration." *Ecosphere* 7, no. 10 (2016): e01471.

Floyd, M. L., W. H. Romme, and D. D. Hanna. "Fire History and Vegetation Pattern in Mesa Verde National Park, Colorado, USA." *Ecological Applications* 10 (2000): 1666–80.

Formichella, Charlotte, Kurt Fristrup, Damon Joyce, Emma Lynch, and Ericka Pilcher. "Yosemite National Park Acoustic Monitoring Report, 2005 & 2006." Accessed August 14, 2019. https://irma.nps.gov/DataStore/DownloadFile/432477.

Forister, M. L., A. C. McCall, J. J. Sanders, J. A. Fordyce, J. H. Thorne, J. O'Brien, D. P. Waetjend, and A. M. Shapiro. "Compounded Effects of Climate Change and Habitat Alteration Shift Patterns of Butterfly Diversity." *Proceedings of the National Academy of Sciences USA* 107, no. 5 (2010): 2088–92.

Garrett, Lisa, Mackenzie Jeffress, Mike Britten, Clinton Epps, Chris Ray, and Susan Wolff. "Pikas in Peril: Multiregional Vulnerability Assessment of a Climate-Sensitive Sentinel Species." *Park Science* 28 , no. 2 (Summer 2011): 9–13.

Gergel, D. R., B. Nijssen, J. T. Abatzoglou, D. P. Lettenmaier, and M. R. Stumbaugh. "Effects of Climate Change on Snowpack and Fire Potential in the Western USA." *Climate Change* 141 (2017): 287–99.

Giersch, J. Joseph, Scott Hotaling, Ryan P. Kovach, Leslie A. Jones, and Clint C. Muhlfeld. "Climate-Induced Glacier and Snow Loss Imperils Alpine Stream Insects." *Global Change Biology* 23, no. 7 (July 2017): 2577–89.

Glacier Park Foundation. "The Fires of 2003: A Synopsis." *The Inside Trail* XVII, no. 1 (Winter 2004): 3.

———. "Glacier Park Fires of 2003." *The Inside Trail* XVII, no. 1 (Winter 2004): 2.

Glazner, Allen F., and Greg M. Stock. *Geology Underfoot in Yosemite National Park*. Missoula, MT: Mountain Press Publishing Company, 2010.

Gonzalez, P., M. Garfin, D. D. Breshears, K. M. Brooks, H. E. Brown, E. H. Elias, A. Gunasekara, N. Huntly, J. K. Maldonado, N. J. Mantua, H. G. Margolis, S. McAfee, B. R. Middleton, and B. H. Udall. "Southwest." In *Impacts, Risks, and Adaptation in the United States: Fourth National Climate Assessment*, Vol. II, edited by D. R. Reidmiller, C. W. Avery, D. R. Easterling, K. E. Kunkel, K. L. M. Lewis, T. K. Maycock, and B. C. Stewart, 1101–84. US Global Change Research Program: Washington, DC, 2018.

Gonzalez, Patrick, Fuyao Wang, Michael Notaro, Daniel J. Vimont, and John W. Williams. "Disproportionate Magnitude of Climate Change in United States National Parks." *Environmental Research Letters* 13, no. 10 (September 2018): 104001.

Graumlich, L. J. "A 1000-Year Record of Temperature and Precipitation in the Sierra Nevada." *Quaternary Research* 39 (1993): 249–55.

Griffin, D., and K. J. Anchukaitis. "How Unusual Is the 2012–2014 California Drought?" *Geophysical Research Letters* 41 (2014): 9017–23.

Grinnell, Joseph, and Tracy Irwin Storer. *Animal Life in the Yosemite: An Account of the Mammals, Birds, Reptiles, and Amphibians in a Cross-Section of the Sierra Nevada*. Berkeley: University of California Press, 1924.

Gross, John E., Michael Tercek, Kevin Guay, Marian Talbert, Tony Chang, Ann Rodman, David Thoma, Patrick Jantz, and Jeffrey T. Morisette. "Historical and Projected Climates to Support Climate Adaptation across the Northern Rocky Mountains." In *Climate Change in Wildlands: Pioneering Approaches to Science and Management*, edited by Andrew J. Hansen, William B Monahan, David M. Theobald, and S. Thomas Olliff, 55–77. Washington, DC: Island Press, 2016.

Haines, Aubrey L. *The Yellowstone Story: A History of Our First National Park*. Rev. ed. Vols. 1–2. Yellowstone National Park, WY: Yellowstone Association for Natural Science, History & Education, Inc., in cooperation with the University Press of Colorado, 1996.

Hall, M. H. P., and D. B. Fagre. "Modeled Climate-Induced Glacier Change in Glacier National Park, 1850–2100." *Bioscience* 53, no. 2 (2003): 131–40.

Halofsky, J. S., D. C. Donato, J. F. Franklin, J. E. Halofsky, D. L. Peterson, and B. J. Harvey. "The Nature of the Beast: Examining Climate Adaptation Options in Forests with Stand-Replacing Fire Regimes." *Ecosphere* 9, no. 3 (2018): e02140.

Hansen, Andrew J., and Linda B. Phillips. "Potential Impacts of Climate Change on Tree Species and Biome Types in the United States Northern Rocky Mountains." In *Climate Change in Wildlands: Pioneering Approaches to Science and Management*, edited by Andrew J Hansen, William B Monahan, David M. Theobald, and S. Thomas Olliff, 174–89. Washington, DC: Island Press, 2016.

Hansen, Andrew, Nate Piekielek, Tony Chang, and Linda Phillips. "Changing Climate Suitability for Forests in Yellowstone and the Rocky Mountains." *Yellowstone Science* 23, no. 1 (2015): 36–43.

Hansen, Warren, Lisa Bate, Steve Gniadek, and Creagh Breuner. "Influence of Streamflow on Reproductive Success in a Harlequin Duck (*Histrionicus histrionicus*) Population in the Rocky Mountains." *Waterbirds* 42, no. 4 (December 2019): 411–42.

Hansen, Winslow D., Kristin H. Braziunas, Werner Rammer, Rupert Seidl, and Monica G. Turner. "It Takes a Few to Tango: Changing Climate and Fire Regimes Can Cause Regeneration Failure of Two Subalpine Conifers." *Ecology* 99, no. 4 (April 2018): 966–77.

Hansen, Winslow D., William H. Romme, Aisha Ba, and Monica G. Turner. "Shifting Ecological Filters Mediate Postfire Expansion of Seedling Aspen (*Populus tremuloides*) in Yellowstone." *Forest Ecology and Management* 362 (2016): 218–30.

Hansen, Winslow D., and Monica G. Turner. "Origins of Abrupt Change? Post-fire Subalpine Conifer Regeneration Declines Nonlinearly with Warming and Drying." *Ecological Monographs* 89, no. 1 (February 2019): e01340.

Harvey, Brian J., Daniel C. Donato, and Monica G. Turner. "Burn Me Twice, Shame on Who? Interactions between Successive Forest Fires across a Temperate Mountain Region." *Ecology* 97, no. 9 (September 2016): 2272–82.

———. "High and Dry: Post-fire Tree Seedling Establishment in Subalpine Forests Decreases with Post-fire Drought and Large Stand-Replacing Burn Patches." *Global Ecology and Biogeography* 25, no. 6 (June 2016): 655–69.

Henningsen, John C., Amy L. Williams, Cynthia M. Tate, Steve A. Kilpatrick, and W. David Walter. "Distribution and Prevalence of *Elaeophora schneideri* in Moose in Wyoming." *Alces* 48 (2012): 35–44.

Hereford, R., R. H. Webb, and C. I. Longpre. "Precipitation History and Ecosystem Response to Multidecadal Precipitation Variability in the Mojave Desert Region, 1893–2001." *Journal of Arid Environments* 67 (2006): 13–34.

Hiza, Margaret M. "The Geologic History of the Absaroka Volcanic Province." *Yellowstone Science* 6, no. 2 (Spring 1998): 2–7.

Iknayan, K. J., and S. R. Beissinger. "Collapse of a Desert Bird Community over the Past Century Driven by Climate Change." *Proceedings of the National Academy of Sciences USA* 115 (2018): 8597–602.

Jepson, M., T. Clabough, C. Caudill, and R. Qualls. "An Evaluation of Temperature and Precipitation Data for Parks of the Mojave Desert Network." Natural Resource Report NPS/MOJN/NRR—2016/1339. Fort Collins, CO: NPS, 2016. Accessed June 17, 2018. https://www.uidaho.edu/-/media/UIdaho-Responsive/Files/cnr/FERL/technical-reports/2016/2016-NPS-MOJN.pdf.

Jones, L. A., C. C. Muhlfeld, and L. A. Marshall. "Projected Warming Portends Seasonal Shifts of Stream Temperatures in the Crown of the Continent Ecosystem, USA and Canada." *Climatic Change* 144, no. 4 (2017): 641–55.

Keane, R. E., L. M. Holsinger, M. F. Mahalovic, and D. F. Tomback. *Restoring Whitebark Pine Ecosystems in the Face of Climate Change.* General Technical Report RMRS-GTR-361. Fort Collins, CO: US Department of Agriculture, Forest Service, Rocky Mountain Research Station, 2017.

Larson, R. P., J. M. Byrne, D. L. Johnson, M. G. Letts, and S. W. Kienzle. "Modelling Climate Change Impacts on Spring Runoff for the Rocky Mountains of Montana and Alberta I: Model Development, Calibration and Historical Analysis." *Canadian Water Resources* 36 (2011): 17–34.

Lenart, Melanie. *Global Warming in the Southwest Projections, Observations and Impacts.* Tucson: University of Arizona Press, 2007.

Lesica, Peter. "Arctic-Alpine Plants Decline over Two Decades in Glacier National Park, Montana, U.S.A." *Arctic, Antarctic, and Alpine Research* 46, no. 2 (2014): 327–32.

Liang, S., M. D. Hurteau, and A. L. Westerling. "Response of Sierra Nevada Forests to Projected Climate-Wildfire Interactions." *Global Change Biology* 23 (2017): 2016–30.

Littell, J. S., E. E. Oneil, D. McKenzie, J. A. Hicke, J. A. Lutz, R. A. Norheim, and M. M. Elsner. "Forest Ecosystems, Disturbance, and Climatic Change in Washington State, USA." *Climate Change* 102 (2010): 129–58.

Luce, C. H., and Z. A. Holden. "Declining Annual Streamflow Distributions in the Pacific Northwest United States, 1948–2006." *Geophysical Research Letters* 36 (2009): L16401.

Lutz, J. A., J. W. van Wagtendonk, and J. F. Franklin. "Climatic Water Deficit, Tree Species Ranges, and Climate Change in Yosemite National Park." *Journal of Biogeography* 37 (2010): 936–50.

Lutz, J. A., J. W. van Wagtendonk, A. E. Thode, J. D. Miller, J. F. Franklin. "Climate, Lightning Ignitions, and Fire Severity in Yosemite National Park, California, USA." *International Journal of Wildland Fire* 18 (2009): 765–74.

MacDonald, Douglas M. *Before Yellowstone: Native American Archaeology in the National Park.* Seattle: University of Washington Press, 2018.

Macfarlane, W. W., J. A. Logan, and W. R. Kern. "An Innovative Aerial Assessment of Greater Yellowstone Ecosystem Mountain Pine Beetle–Caused Whitebark Pine Mortality." *Ecological Applications* 23 (2013): 421–37.

Mantua, N., I. Tohver, and A. Hamlet. "Climate Change Impacts on Streamflow Extremes and Summertime Stream Temperature and Their Possible Consequences for Freshwater Salmon Habitat in Washington State." *Climate Change* 102 (2010): 187–223.

Marlier, Miriam E., Mu Xiao, Ruth Engel, Ben Livneh, John T. Abatzoglou, and Dennis P. Lettenmaier. "The 2015 Drought in Washington State: A Harbinger of Things to Come?" *Environmental Research Letters* 12, no. 11 (November 2017): 114008.

Martin, R. E., G. P. Asner, E. Francis, A. Ambrose, W. Baxter, A. J. Das, N. Vaughn, T. Paz-Kagan, T. Dawson, K. Nydick, and N. L. Stephenson. "Remote Analysis of Canopy Water Content in Giant Sequoias (*Sequoiadendron giganteum*) during Drought." *Forest Ecology and Management* 419–20 (2018): 279–90.

Marzeion, B., J. G. Cogley, K. Richter, and D. Parkes. "Attribution of Global Glacier Mass Loss to Anthropogenic and Natural Causes." *Science* 345 (2014): 919–21.

McAuliffe, J. R., and E. P. Hamerlynck. "Perennial Plant Mortality in the Sonoran and Mojave Deserts in Response to Severe, Multi-year Drought." *Journal of Arid Environments* 74 (2010): 885–96.

McCabe, G. J., D. M. Wolock, G. T. Pederson, C. A. Woodhouse, and S. McAfee. "Evidence That Recent Warming Is Reducing Upper Colorado River Flows." *Earth Interactions* 21, no. 10 (2017): 1–14.

McIntyre, P. J., J. H. Thorne, C. R. Dolanc, A. L. Flint, L. E. Flint, M. Kelly, and D. D. Ackerly. "Twentieth-Century Shifts in Forest Structure in California: Denser Forests, Smaller Trees, and Increased Dominance of Oaks." *Proceedings of the National Academy of Sciences* 112 (2015): 1458–63.

McKinney, Shawn T., Carl E. Fiedler, and Diana F. Tomback. "Invasive Pathogen Threatens Bird-Pine Mutualism: Implications for Sustaining a High-Elevation Ecosystem." *Ecological Applications* 19, no. 3 (2009): 597–607.

McNulty, Tim. *Olympic National Park: A Natural History*. 4th ed. Seattle: University of Washington Press, 2018.

Millar, C. I., R. D. Westfall, D. L. Delany, J. C. King, and L. J. Graumlich. "Response of Subalpine Conifers in the Sierra Nevada, California, USA, to 20th-Century Warming and Decadal Climate Variability." *Arctic, Antarctic, and Alpine Research* 36 (2004): 181–200.

Mills, L. Scott, Marketa Zimova, Jared Oyler, Steven Running, John T. Abatzoglou, and Paul M. Lukacs. "Camouflage Mismatch in Seasonal Coat Color Due to Decreased Snow Duration." *Proceedings of the National Academy of Sciences* 110 , no. 18 (April 2013): 7360–65.

Millspaugh, Sarah H., Cathy Whitlock, and Patrick J. Bartlein. "Variations in Fire Frequency and Climate over the Past 17000 Yr in Central Yellowstone National Park." *Geology* 28, no. 3 (2000): 211–14.

Monahan, William B., and Nicholas A. Fisichelli. "Climate Exposure of US National Parks in a New Era of Change." *PLoS ONE* 9, no. 7 (2014): e101302.

———. "Recent Climate Change Exposure of Glacier National Park." July 2014. Accessed January 4, 2019. https://irma.nps.gov/DataStore/DownloadFile/497470.

———. "Recent Climate Change Exposure of Grand Canyon National Park." July 2014. Accessed December 19, 2018. https://irma.nps.gov/DataStore/DownloadFile/496801.

———. "Recent Climate Change Exposure of Olympic National Park." July 2014. Accessed December 29, 2018. https://irma.nps.gov/DataStore/DownloadFile/497286.

———. "Recent Climate Change Exposure of Yellowstone National Park." July 2014. Accessed March 10, 2019. https://irma.nps.gov/DataStore/DownloadFile/497075.

———. "Recent Climate Change Exposure of Yosemite National Park." July 2014. Accessed August 31, 2019. https://irma.nps.gov/DataStore/DownloadFile/497022.

Moore, P. E., O. Alvarez, S. T. McKinney, W. Li, M. L. Brooks, and Q. Guo. "Climate Change and Tree-Line Ecosystems in the Sierra Nevada: Habitat Suitability Modelling to Inform High-Elevation Forest Dynamics Monitoring." Natural Resource Report NPS/SIEN/NRR 2017/1476. Fort Collins, Colorado: NPS, 2017.

Morelli, T. L., A. B. Smith, C. R. Kastely, I. Mastroserio, C. Moritz, and S. R. Beissinger. "Anthropogenic Refugia Ameliorate the Severe Climate-Related Decline of a Montane Mammal along Its Trailing Edge." *Proceedings of the Royal Society B: Biological Sciences* 279 (2012): 4279–86.

Moritz, C., J. L. Patton, C. J. Conroy, J. L. Parra, and G. C. White. "Impact of a Century of Climate Change on Small-Mammal Communities in Yosemite National Park, USA." *Science* 322 (2008): 261–64.

Mote, P. W., S. Li, D. P. Lettenmaier, M. Xiao, and R. Engel. "Dramatic Declines in Snowpack in the Western US." *Climate and Atmospheric Science* 1, no. 1 (2018): 2.

Mote, P. W., D. E. Rupp, S. Li, D. J. Sharp, F. Otto, P. F. Uhe, M. Xiao, D. P. Lettenmaier, H. Cullen, and M. R. Allen. "Perspectives on the Causes of Exceptionally Low 2015 Snowpack in the Western United States." *Geophysical Research Letters* 43, no. 10 (2016): 980–88.

Moyer-Horner, L., E. A. Beever, D. H. Johnson, M. Biel, and J. Belt. "Predictors of Current and Longer-Term Patterns of Abundance of American Pikas (*Ochotona princeps*) across a Leading-Edge Protected Area." *PLoS ONE* 11, no. 11 (2016): e0167051.

Muhlfeld, Clint C., J. Joseph Giersch, F. Richard Hauer, Gregory T. Pederson, Gordon Luikart, Douglas P. Peterson, Christopher C. Downs, Daniel B. Fagre. "Climate Change Links Fate of Glaciers and an Endemic Alpine Invertebrate." *Climatic Change* 106, no. 2 (May 2011): 337–45.

Nabokov, Peter, and Lawrence Loendorf. *Restoring a Presence: American Indians and Yellowstone National Park*. Norman: University of Oklahoma Press, 2004.

Nakawatase, J. M., and D. L. Peterson. "Spatial Variability in Forest Growth—Climate Relationships in the Olympic Mountains, Washington." *Canadian Journal of Forest Research* 36 (2006): 77–91.

Nelson, Kellen N., Monica G. Turner, William H. Romme, and Daniel B. Tinker. "Landscape Variation in Tree Regeneration and Snag Fall Drive Fuel Loads in 24-Year Old Post-fire Lodgepole Pine Forests." *Ecological Applications* 26, no. 8 (2016): 2424–38.

———. "Simulated Fire Behaviour in Young, Postfire Lodgepole Pine Forests." *International Journal of Wildland Fire* 26 (2017): 852–65.

Nesmith, Jonathan C. B., Micah Wright, Erik S. Jules, and Shawn T. McKinney. "Whitebark and Foxtail Pine in Yosemite, Sequoia, and Kings Canyon National Parks: Initial Assessment of Stand Structure and Condition." *Forests* 10 (2019): 35.

Nydick, K. R., A. Ambrose, G. P. Asner, W. Baxter, A. J. Das, R. E. Martin, T. Paz-Kagan, and N. L. Stephenson. "The Leaf to Landscape Project: Empirical Mapping of Climate Change Vulnerability for Giant Sequoia in the Southern Sierra Nevada, California." *Forest Ecology and Management* 419–20 (2018): 249–56.

O'Neel, S., C. McNeil, L. C. Sass, C. Florentine, E. H. Baker, E. Peitzsch, D. McGrath, A. G. Fountain, and D. Fagre. "Reanalysis of the US Geological Survey Benchmark Glaciers: Long-Term Insight into Climate Forcing of Glacier Mass Balance." *Journal of Glaciology* 65, no. 253 (2019): 850–66.

Oreskes, Naomi. "Beyond the Ivory Tower: The Scientific Consensus on Climate Change." *Science* 306, no. 5702 (December 2004): 1686.

Pachauri, R. K., and L. A. Meyer, eds.. *Climate Change 2014: Synthesis Report: Contribution of Working Groups I, II and III to the Fifth Assessment Report of the Intergovernmental Panel on Climate Change.* Geneva, Switzerland: Intergovernmental Panel on Climate Change, 2015.

Paz-Kagan, T., N. R. Vaughn, R. E. Martin, P. Brodrick, N. L. Stephenson, A. J. Das, K. R. Nydick, and G. P. Asner. "Landscape-Scale Variation in Canopy Water Content of Giant Sequoias during Drought." *Forest Ecology and Management* 419–20 (2018): 291–304.

Peacock, Synte. "Projected 21st Century Climate Change for Wolverine Habitats within the Contiguous United States." *Environmental Research Letters* 6, no. 1 (January 2011): 25.

Pederson, G. T., J. L. Betancourt, and G. J. McCabe. "Regional Patterns and Proximal Causes of the Recent Snowpack Decline in the Rocky Mountains, US." *Geophysical Research Letters* 40 (2013): 1811–16, 2013.

Pederson, Gregory T., Lisa J. Graumlich, Daniel B. Fagre, Todd Kipfer, and Clint C. Muhlfeld. "A Century of Climate and Ecosystem Change in Western

Montana: What Do Temperature Trends Portend?" *Climatic Change* 98, nos. 1–2 (January 2010): 133–54.

Pederson, Gregory T., S. T. Gray, T. Ault, W. Marsh, D. B. Fagre, A. G. Bunn, C. A. Woodhouse, and L. J. Graumlich. "Climatic Controls on the Snowmelt Hydrology of the Northern Rocky Mountains." *Journal of Climate* 24, no. 6 (2011): 1666–87.

Pederson, Gregory T., Stephen T. Gray, C. A. Woodhouse, Julio L. Betancourt, Daniel B. Fagre, Jeremy S. Littell, Emma Watson, B. H. Luckman, and Lisa J. Graumlich. "The Unusual Nature of Recent Snowpack Declines in the North American Cordillera." *Science* 333, no. 6040 (July 2011): 332–35.

Peterson, D. W., D. L. Peterson, and G. J. Ettl. "Growth Responses of Subalpine Fir to Climatic Variability in the Pacific Northwest." *Canadian Journal of Forest Research* 32 (2002): 1503–17.

Powell, James L. "The Consensus on Anthropogenic Global Warming Matters." *Bulletin of Science, Technology & Society* 36, no. 3 (2017): 157–63.

———. *Dead Pool: Lake Powell, Global Warming, and the Future of Water in the West*. Berkeley: University of California Press, 2008.

Raffa, Kenneth F., Erinn N. Powell, and Philip A. Townsend. "Temperature-Driven Range Expansion of an Irruptive Insect Heightened by Weakly Coevolved Plant Defenses." *Proceedings of the National Academy of Sciences* 110, no. 6 (February 2013): 2193–98.

Rasmussen, L. A., and H. Conway. "Estimating South Cascade Glacier (Washington, USA) Mass Balance from a Distant Radiosonde and Comparison with Blue Glacier." *Journal of Glaciology* 47, no. 159 (2001): 579–88.

Ray, A. M., W. R. Gould, B. R. Hossack, A. J. Sepulveda, D. P. Thoma, D. A. Patla, R. Daley, and R. Al-Chokhachy. "Influence of Climate Drivers on Colonization and Extinction Dynamics of Wetland-Dependent Species." *Ecosphere* 7, no. 7 (2016): e01409.

Ray, Andrew, Adam Sepulveda, Blake Hossack, Debra Patla, David Thoma, Robert Al-Chokhachy, and Andrea Litt. "Monitoring Greater Yellowstone Ecosystem Wetlands: Can Long-Term Monitoring Help Us Understand Their Future?" *Yellowstone Science* 23, no. 1 (2015): 44–52.

Ray, C., J. F. Saracco, M. L. Holmgren, R. L. Wilkerson, R. B. Siegel, K. J. Jenkins, J. I. Ransom, P. J. Happe, J. R. Boetsch, and M. H. Huff. "Recent Stability of Resident and Migratory Landbird Populations in National Parks of the Pacific Northwest." *Ecosphere* 8, no. 7 (2017): e01902.

Redmond, M. D., F. Forcella, and N. N. Barger. "Declines in Pinyon Pine Cone Production Associated with Regional Warming." *Ecosphere* 3 (2012): 120.

Rehfeldt, G. E., N. L. Crookston, M. V. Warwell, and J. S. Evans. "Empirical

Analyses of Plant-Climate Relationships for the Western United States." *International Journal of Plant Sciences* 167, no. 6 (2006): 1123–50.
Reisner, Marc. *Cadillac Desert: American West and Its Disappearing Water*. Rev. ed. New York: Penguin, 1993.
Riddell, Eric A., Kelly J. Iknayan, Blair O. Wolf, Barry Sinervo, and Steven R. Beissinger. "Cooling Requirements Fueled the Collapse of a Desert Bird Community from Climate Change." *Proceedings of the National Academy of Sciences* 116, no. 43 (October 2019): 21609–15.
Riedel, Jon L., Steve Wilson, William Baccus, Michael Larrabee, T. J. Fudge, and Andrew G. Fountain. "Glacier Status and Contribution to Streamflow in the Olympic Mountains, Washington, USA." *Journal of Glaciology* 61, no. 225 (2015): 8–16.
Righter, Robert. *The Battle Over Hetch Hetchy: America's Most Controversial Dam and the Birth of Modern Environmentalism*. New York: Oxford University Press, 2005.
Robeson, Scott M. "Revisiting the Recent California Drought as an Extreme Value." *Geophysical Research Letters* 42 (2015): 6771–79.
Roe, G. H., M. B. Baker, and F. Herla. "Centennial Glacier Retreat as Categorical Evidence of Regional Climate Change." *Nature Geoscience* 10 (2107): 95–99.
Romme, W. H., C. D. Allen, J. D. Bailey, W. L. Baker, B. T. Bestelmeyer, P. M. Brown, K. S. Eisenhart, M. L. Floyd, D. W. Huffman, B. F. Jacobs, R. F. Miller, E. H. Muldavin, T. W. Swetnam, R. J. Tausch, and P. J. Weisberg. "Historical and Modern Disturbance Regimes, Stand Structures, and Landscape Dynamics in Piñon-Juniper Vegetation of the Western United States." *Rangeland Ecology and Management* 62 (2009): 203–22.
Romme, William H. "Fire and Landscape Diversity in Subalpine Forests of Yellowstone National Park." *Ecological Monographs* 52, no. 2 (February 1982): 199–221.
Romme, William H., Mark S. Boyce, Robert Gresswell, Evelyn H. Merrill, G. Wayne Minshall, Cathy Whitlock, and Monica G. Turner. "Twenty Years after the 1988 Yellowstone Fires: Lessons about Disturbance and Ecosystems." *Ecosystems* 14, no. 7 (2011): 1196–215.
Romme, William H., Monica G. Turner, R. H. Gardner, W. W. Hargrove, G. A. Tuskan, Donald G. Despain, and Roy A. Renkin. "A Rare Episode of Sexual Reproduction in Aspen (*Populus tremuloides Michx.*) following the 1988 Yellowstone Fires." *Natural Areas Journal* 17 (1997): 17–25.
Romme, William H., Monica G. Turner, G. A. Tuskan, and R. A. Reed. "Establishment, Persistence, and Growth of Aspen (*Populus tremuloides*) Seedlings in Yellowstone National Park." *Ecology* 86, no. 2 (2005): 404–18.

Romme, William H., Monica G. Turner, Linda L. Wallace, and J. Walker. "Aspen, Elk, and Fire in Northern Yellowstone National Park." *Ecology* 76 (1995): 2097–106.

Romme, William H., Timothy G. Whitby, Daniel B. Tinker, and Monica G. Turner. "Deterministic and Stochastic Processes Lead to Divergence in Plant Communities 25 Years after the 1988 Yellowstone Fires." *Ecological Monographs* 86, no. 3 (2016): 327–51.

Roush, W., J. S. Munroe, D. B. Fagre. "Development of a Spatial Analysis Method Using Ground-Based Repeat Photography to Detect Changes in the Alpine Treeline Ecotone, Glacier National Park, Montana, U.S.A." *Arctic, Antarctic, and Alpine Research* 39, no. 2 (2007): 297–308.

Rowe, K. C., K. M. Rowe, M. W. Tingley, M. S. Koo, J. L. Patton, C. Conroy, J. D. Perrine, S. M. Beissinger, and C. Moritz. "Spatially Heterogeneous Impact of Climate Change on Small Mammals of Montane California." *Proceedings of the Royal Society B: Biological Sciences* 282, no. 1799 (2015): 20141857.

Runte, Al. *Yosemite: The Embattled Wilderness*. Lincoln: University of Nebraska Press, 1990.

Rupp, David E., and Sihan Li. "Less Warming Projected during Heavy Winter Precipitation in the Cascades and Sierra Nevada." *International Journal of Climatology* 37, no. 10 (August 2017): 3984–90.

Salzer, Matthew W., and Kurt F. Kipfmueller. "Reconstructed Temperature and Precipitation on a Millennial Timescale from Tree-Rings in the Southern Colorado Plateau, USA." *Climatic Change* 70 (2005): 465–87.

Sax, Joseph. *Mountains without Handrails: Reflections on the National Parks*. Ann Arbor: University of Michigan Press, 1980.

Schmidt, Jeremy. *Grand Canyon: A Natural History Guide*. New York: Houghton Mifflin, 1993.

Schoennagel, T., Monica G. Turner, and William H. Romme. "The Influence of Fire Interval and Serotiny on Postfire Lodgepole Pine Density in Yellowstone National Park." *Ecology* 84 (2003): 1967–78.

Schwartz, M. W., N. Butt, C. R. Dolanc, A. Holguin, M. A. Moritz, M. P. North, H. D. Safford, N. L. Stephenson, J. H. Thorne, and P. J. van Mantgem. "Increasing Elevation of Fire in the Sierra Nevada and Implications for Forest Change." *Ecosphere* 6, no. 7 (2015): 121.

Seager, R., M. Hoerling, S. Schubert, H. Wang, B. Lyon, A. Kumar, J. Nakamura, and N. Henderson. "Causes of the 2011 to 2014 California Drought." *Journal of Climatology* 28 (2015): 6997–7024.

Seifert, D., E. Chatelain, C. Lee, Z. Seligman, D. Evans, H. Fisk, and P. Maus. "Monitoring Alpine Climate Change in the Beartooth Mountains of the Custer National Forest." USFS report RSAC-0115-RPT1. Salt Lake City, UT:

USFS, 2009. Accessed March 31, 2019. https://www.fs.usda.gov/Internet/FSE_DOCUMENTS/stelprd3834828.pdf.

Shanahan, E., K. M. Irvine, D. Thoma, S. Wilmoth, A. Ray, K. Legg, and H. Shovic. "Whitebark Pine Mortality Related to White Pine Blister Rust, Mountain Pine Beetle Outbreak, and Water Availability." *Ecosphere* 7, no. 12 (2016): e01610.

Shaw, J. D., B. E. Steed, and L. T. DeBalder. "Forest Inventory and Analysis (FIA) Annual Inventory Answers the Question: What Is Happening to Pinyon-Juniper Woodlands?" *Journal of Forestry* 103 (2005): 280–85.

Sheehan T., D. Bachelet, and K. Ferschweiler. "Projected Major Fire and Vegetation Changes in the Pacific Northwest of the Conterminous United States under Selected CMIP5 Climate Futures." *Ecological Modeling* 317 (2015): 16–29.

Shukla, S., M. Safeeq, A. AghaKouchak, K. Guan, and C. Funk. "Temperature Impacts on the Water Year 2014 Drought in California." *Geophysical Research Letters* 42 (2015): 4384–93.

Smedes, Harry W., and Harold J. Prostka. *Stratigraphic Framework of the Absaroka Volcanic Supergroup in the Yellowstone National Park Region.* US Geological Survey Professional Paper 729-C. Washington, DC: US Government Printing Office, 1972.

Stephenson, Nathan L., Adrian J. Das, Nicholas J. Ampersee, Beverly M. Bulaon, and Julie L. Yee. "Which Trees Die during Drought? The Key Role of Insect Host-Tree Selection." *Journal of Ecology* 107, no. 3 (2019): 1–19.

Stephenson, N. L., A. J. Das, N. J. Ampersee, K. G. Cahill, A. C. Caprio, J. E. Sanders, and A. P. Williams. "Patterns and Correlates of Giant Sequoia Foliage Dieback during California's 2012–2016 Hotter Drought." *Forest Ecology and Management* 419–20 (2018): 268–78.

Stevens-Rumann, Camille S., Kerry B. Kemp, Philip E. Higuera, Brian J. Harvey, Monica T. Rother, Daniel C. Donato, Penelope Morgan, and Thomas T. Veblen. "Evidence for Declining Forest Resilience to Wildfires under Climate Change." *Ecology Letters* 21, no. 2 (February 2018): 243–52.

Stewart, Joseph A. E., John D. Perrine, Lyle B. Nichols, James H. Thorne, Constance I. Millar, Kenneth E. Goehring, Cody P. Massing, and David H. Wright. "Revisiting the Past to Foretell the Future: Summer Temperature and Habitat Area Predict Pika Extirpations in California." *Journal of Biogeography* 42 (2015): 880–90.

Stine, Scott. "Extreme and Persistent Drought in California and Patagonia during Mediaeval Time." *Nature* 369 (1994): 546–49.

Stock, Greg. "Vanishing Ice, Vanishing History." *Mountain Views* 9, no. 2 (December 2015): 22–26.

Swain, Daniel L., Daniel E. Horton, Deepti Singh, and Noah S. Diffenbaugh.

"Trends in Atmospheric Patterns Conducive to Seasonal Precipitation and Temperature Extremes in California." *Science Advances* 2, no. 4 (April 2016): e1501344.

Sweet, L. C., T. Green, J. G. C. Heintz, N. Frakes, N. Graver, J. S. Rangitsch, J. E. Rodgers, S. Heacox, and C. W. Barrows. "Congruence between Future Distribution Models and Empirical Data for an Iconic Species at Joshua Tree National Park." *Ecosphere* 10, no. 6 (2019): e02763.

Syphard, A. D., J. E. Keeley, and J. T. Abatzoglou. "Trends and Drivers of Fire Activity Vary across California Aridland Ecosystems." *Journal of Arid Environments* 144 (2017): 110–22.

Tercek, Mike. "Nowcasting & Forecasting Fire Severity in Yellowstone." *Yellowstone Science* 27, no. 1 (April 2019): 27–33.

———. "A Seemingly Small Change in Average Temperature Can Have Big Effects." *Yellowstone Science* 23 (2015): 70–71.

Tercek, Michael, and Ann Rodman. "Forecasts of 21st Century Snowpack and Implications for Snowmobile and Snowcoach Use in Yellowstone National Park." *PLoS ONE* 11, no. 7 (2016): e015921.

Tercek, Mike, Ann Rodman, and David Thoma. "Trends in Yellowstone's Snowpack." *Yellowstone Science* 23, no. 1 (2015): 20–27.

Thoma, David, J. Norris, and P. Lauck. "Satellite-Based Vegetation Condition and Phenology, and Snow Cover Extent Monitoring Protocol for the Northern and Southern Colorado Plateau Networks." Natural Resource Report NPS/SCPN/NRR—2017/1533. Fort Collins, CO: NPS, 2017.

Thoma, David, Ann Rodman, and Mike Tercek. "Water in the Balance: Interpreting Climate Change Impacts Using a Water Balance Model." *Yellowstone Science* 23, no. 1 (2015): 29–35.

Thoma, David P., Seth M. Munson, Ann W. Rodman, Roy Renkin, Heidi M. Anderson, and Stefanie D. Wacker. "Patterns of Primary Production & Ecological Drought in Yellowstone." *Yellowstone Science* 27, no. 1 (April 2019): 34–39.

Thoma, D. P., S. M. Munson, and D. L. Witwicki. "Landscape Pivot Points and Responses to Water Balance in National Parks of the Southwest U.S." *Journal of Applied Ecology* 56 (2019): 157–67.

Thoma, D. P., E. K. Shanahan, and K. M. Irvine. "Climatic Correlates of White Pine Blister Rust Infection in Whitebark Pine in the Greater Yellowstone Ecosystem." *Forests* 10 (2019): 666.

Tingley, Morgan W., Michelle Koo, Craig Moritz, Andrew Rush, and Steven R. Beissinger. "The Push and Pull of Climate Change Causes Heterogeneous Shifts in Avian Elevational Ranges." *Global Change Biology* 18, no. 11 (2012): 3279–90.

Turner, Monica G., Kristin H. Braziunas, Winslow D. Hansen, and Brian J.

Harvey. "Short-Interval Severe Fire Erodes the Resilience of Subalpine Lodgepole Pine Forests." *Proceedings of the National Academy of Sciences* 116, no. 23 (June 2019): 11319–28.

Turner, Monica G., Daniel C. Donato, Winslow D. Hansen, Brian J. Harvey, William H. Romme, and A. LeRoy Westerling. "Climate Change and Novel Disturbance Regimes in National Park Landscapes." In *Science, Conservation, and National Parks*, edited by Steven R. Beissinger, David D. Ackerly, Holly Doremus, and Gary E. Machlis, 77–101. Chicago: University of Chicago Press, 2017.

Turner, Monica G., William H. Romme, and Daniel B. Tinker. "Surprises and Lessons from the 1988 Yellowstone Fires." *Frontiers in Ecology and the Environment* 1, no. 7 (September 2003): 351–58.

Udall, Bradley, and Jonathan Overpeck. "The Twenty-First Century Colorado River Hot Drought and Implications for the Future." *Water Resources Research*, 53, no. 3 (2017): 2404–18.

van Mantgem, Phillip J., Nathan L. Stephenson, John C. Byrne, Lori D. Daniels, Jerry F. Franklin, Peter Z. Fulé, Mark E. Harmon, Andrew J. Larson, Jeremy M. Smith, Alan H. Taylor, and Thomas T. Veblen. "Widespread Increase of Tree Mortality Rates in the Western United States." *Science* 323, no. 5913 (January 2009): 521–24.

Vano, Julie A., Bradley Udall, Daniel R. Cayan, Jonathan T. Overpeck, Levi D. Brekke, Tapash Das, Holly C. Hartmann, Hugo G. Hidalgo, Martin Hoerling, Gregory J. McCabe, Kiyomi Morino, Robert S. Webb, Kevin Werner, and Dennis P. Lettenmaier. "Understanding Uncertainties in Future Colorado River Streamflow." *Bulletin of the American Meteorological Society* 95, no. 1 (2014): 59–78.

Wahrhaftig, Clyde, Greg M. Stock, Reba G. McCracken, Peri Sasnett, and Andrew J. Cyr. "Extent of the Last Glacial Maximum (Tioga) Glaciation in Yosemite National Park and Vicinity, California." US Geological Survey Scientific Investigation. Reston, VA: US Geological Survey, 2019.

Wallace, Linda L., ed. *After the Fires: The Ecology of Change in Yellowstone National Park*. New Haven, CT: Yale University Press, 2004.

Wells, Gail. "Clark's Nutcracker and Whitebark Pine: Can the Birds Help the Embattled High-Country Pine Survive?" *Pacific Northwest Research Station Science Findings* 130 (February 2011). Accessed April 2, 2019. https://www.fs.fed.us/pnw/sciencef/scifi130.pdf.

Weiss, J. L., and J. T. Overpeck. "Is the Sonoran Desert Losing Its Cool?" *Global Change Biology* 11 (2005): 2065–77.

Westerling, Anthony, H. G. Hidalgo, D. R. Cayan, and T. W. Swetnam. "Warming and Earlier Spring Increase Western US Forest Wildfire Activity." *Science* 313, no. 5789 (2006): 940–43.

Westerling, Anthony L. "Increasing Western US Forest Wildfire Activity:

Sensitivity to Changes in the Timing of Spring." *Philosophical Transactions of the Royal Society B* 371, no. 1696 (June 2016): 20150178.
Westerling, Anthony L., Monica G. Turner, Erica A. H. Smithwick, William H. Romme, and Michael G. Ryan. "Continued Warming Could Transform Greater Yellowstone Fire Regimes by Mid-21st Century." *Proceedings of the National Academy of Sciences, USA* 108 (2011): 13165–70.
Whitlock, Cathy, and Steve Hostetler. "Past Warm Periods Provide Vital Benchmarks for Understanding the Future of the Greater Yellowstone Ecosystem." *Yellowstone Science* 27, no. 1 (April 2019): 72–76.
Williams, A. P., C. D. Allen, C. I. Millar, T. W. Swetnam, J. Michaelsen, C. J. Still, and S. W. Leavitt. "Forest Responses to Increasing Aridity and Warmth in the Southwestern United States." *Proceedings of the National Academy of Sciences, USA* 107 (2010): 21289–94.
Williams, A. P., R. Seager, J. T. Abatzoglou, B. I. Cook, J. E. Smerdon, and E. R. Cook. "Contribution of Anthropogenic Warming to California Drought during 2012–2014." *Geophysical Research Letters* 42 (2015): 6819–28.
Woodward, A., E. G. Schreiner, and D. G. Silsbee. "Climate, Geography, and Tree Establishment in Subalpine Meadows of the Olympic: Mountains, Washington, U.S.A." *Arctic and Alpine Research* 27, no. 3 (1995): 217–25.
Yochim, Michael J. "Aboriginal Overkill Overstated: Errors in Charles Kay's Hypothesis." *Human Nature* 12 (2001): 141–67.
———. *Essential Yellowstone: A Landscape of Memory and Wonder*. Helena, MT: Riverbend Publishing, 2019.
———. *Protecting Yellowstone: Science and the Politics of National Park Management*. Albuquerque: University of New Mexico Press, 2013.
———. *A Week in Yellowstone's Thorofare: A Journey through the Remotest Place*. Corvallis: Oregon State University Press, 2016.
———. *Yellowstone and the Snowmobile: Locking Horns over Winter Use*. Lawrence: University Press of Kansas, 2009.
Yochim, Michael J., and William R. Lowry. "Creating Conditions for Policy Change in National Parks." *Environmental Management* 57, no. 5 (2016): 1041–53.

Internet Sources and News Articles

American Cancer Society. "Lifetime Risk of Developing or Dying From Cancer." Accessed March 24, 2018. https://www.cancer.org/cancer/cancer-basics/lifetime-probability-of-developing-or-dying-from-cancer.html.
American Forests. "Champion Trees National Register." Accessed July 20, 2019. https://www.americanforests.org/get-involved/americas-biggest-trees/champion-trees-national-register/.

American Heart Association and American Stroke Association. "Heart Disease and Stroke Statistics—at a Glance." Accessed March 24, 2018. https://www.heart.org/idc/groups/ahamah-public/@wcm/@sop/@smd/documents/downloadable/ucm_470704.pdf.
American Society of Civil Engineers. "Colorado River Aqueduct." Accessed October 18, 2019. https://www.asce.org/project/colorado-river-aqueduct/.
Bloch, Sam. "California Must Abandon 535,000 Acres of Prized Farmland to Meet Water Conservation Goals." *New Food Economy*, February 28, 2019. Accessed October 19, 2019. https://newfoodeconomy.org/california-san-joaquin-valley-farmland-groundwater-aquifer-drought/.
Burwell, Lakota. "Environmental Factors Influencing Fire Behavior in the Olympic Mountains." Humboldt State University capstone project, n.d. Accessed May 3, 2018. https://www.nps.gov/olym/learn/management/fire-history.htm.
California Department of Food and Agriculture. "California Agricultural Production Statistics." Accessed October 9, 2019. https://www.cdfa.ca.gov/statistics.
California Department of Water Resources. "Current Conditions." Accessed October 10, 2019. https://water.ca.gov/News/Current-Conditions.
———. "State Water Project." Accessed October 9, 2019. https://water.ca.gov/Programs/State-Water-Project#.
Cataldo, John, and Becky Smith. "2016 Yellowstone Fires." PowerPoint presentation as part of Northern Rockies Fire Science Network, "Long Duration Fire and Re-Burn Effects in Yellowstone National Park." Accessed May 10, 2019. https://www.nrfirescience.org/event/long-duration-fire-and-re-burn-effects-yellowstone-national-park.
Cornell Lab of Ornithology. "All about Birds: Canyon Wren." Accessed March 28, 2018. https://www.allaboutbirds.org/guide/Canyon_Wren/overview.
Cusick, Daniel. "Rapid Climate Changes Turn North Woods into Moose Graveyard." *Scientific American*, May 18, 2012. Accessed May 24, 2019. https://www.scientificamerican.com/article/rapid-climate-changes-turn-north-woods-into-moose-graveyard/.
Davis, Tony. "Risks to Lake Mead, Colorado River Intensifying Greatly, Federal Officials Say." *Arizona Daily Star*, June 29, 2018. Accessed August 12, 2018. https://tucson.com/news/local/risks-to-lake-mead-colorado-river-intensifying-greatly-federal-officials/article_ed5a3146–29fb-59ba-ab6b-a65d34ba258b.html.
Earthwatch Institute. "Climate Change, Huckleberries, and Grizzly Bears in Montana." Accessed October 11, 2018. https://au.earthwatch.org/Expeditions/Climate-Change-Huckleberries-and-Grizzly-Bears-in-Montana.
Fourth National Climate Assessment. Accessed throughout 2019. https://nca2018.globalchange.gov/.
Glacier National Park Conservancy. "Black Swifts: The Poster Bird for Climate

Change." Accessed October 7, 2018. https://glacier.org/newsblog/project/black-swifts-poster-bird-climate-change-2.
Grand Canyon Trust. "Stopping Grand Canyon Escalade." Accessed July 2, 2018. https://www.grandcanyontrust.org/stopping-grand-canyon-escalade.
Griggs, Gary. "Sea-Level Rise for the Coasts of California, Oregon, and Washington: Past, Present, and Future." Posted on June 22, 2012. YouTube video, 4:42. Accessed December 5, 2019. https://www.youtube.com/watch?v=-gstw44DeSI.
Hegewisch, K. C., and J. T. Abatzoglou. "Future Time Series." Web tool. Northwest Climate Toolbox. Accessed throughout 2018 and 2019. https://climatetoolbox.org/.
Lindsey, Rebecca. "Climate Change: Global Sea Level." Climate.gov, November 19, 2019. Accessed December 6, 2019. https://www.climate.gov/news-features/understanding-climate/climate-change-global-sea-level.
Mariposa County. "Post Ferguson Fire." Accessed November 7, 2019. https://www.yosemite.com/post-ferguson-fire/.
Metcalfe, John. "Watch the Pacific Northwest Change Before Your Eyes." *Bloomberg CityLab*, May 19, 2017. Accessed December 2, 2019. https://www.citylab.com/environment/2017/05/watch-the-pacific-northwest-change-before-your-eyes/527136/.
National Aeronautics and Space Administration (NASA). "Climate Change: How Do We Know?" Accessed February 25, 2020. https://climate.nasa.gov/evidence/.
———. "Global Climate Change: Vital Signs of the Planet." Accessed November 19, 2018. https://climate.nasa.gov/evidence/.
———. "Paleoclimatology: The Oxygen Balance." Accessed February 25, 2020. https://earthobservatory.nasa.gov/features/Paleoclimatology_OxygenBalance.
National Climate Assessment. Accessed throughout 2018. https://nca2014.globalchange.gov.
Natural Resources Defense Council. "California Snowpack and the Drought." Fact sheet. April 2014. Accessed October 5, 2019. https://www.nrdc.org/sites/default/files/ca-snowpack-and-drought-FS.pdf.
NOAA Climate.gov. "Climate Models." Accessed July 26, 2018. https://www.climate.gov/maps-data/primer/climate-models.
NPS. "Birds and Climate Change: Grand Canyon National Park." Accessed January 12, 2020. https://www.nps.gov/subjects/climatechange/upload/GRCA_2018_Birds_-_CC_508Compliant.pdf.
———. "Parsons Memorial Lodge." Accessed August 29, 2019. https://www.nps.gov/parkhistory/online_books/harrison/harrison10.htm.
NPS, Glacier National Park. "Climate Change," "Fire-Fueled Finds," "Fire History," "Geology," "Howe Ridge Fire Update, August 14 11 AM," "Ice Patch

Archeology Resource Brief," "National Park Service Releases Review of Fire at Glacier National Park's Sperry Chalet," "Nature & Science: Flora, Fauna, and Other Life Forms," "Pikas Resource Brief," "Plants," and "Wildfire History at Glacier National Park." Accessed August–October 2018. https://www.nps.gov/glac/.

NPS, Grand Canyon National Park. "Geologic Activity," "Geologic Formations," "Park Statistics," "Plants: Grand Canyon Species List," "Mammals," and "Reptiles." Accessed May 25–August 6, 2018. https://www.nps.gov/grca/learn/.

NPS, Grand Teton National Park. "Glacier Monitoring." Accessed March 30, 2019. https://www.nps.gov/grte/learn/nature/glaciermonitoring.htm.

NPS, Inventory and Monitoring Program (North Coast and Cascades). "Vital Signs Overviews." Accessed November 2018. https://static1.squarespace.com/static/5199bbc2e4b00dab067c9432/t/51c14a34e4b0e0c8e6cf4ab1/1371621940869/JAndrews_NPS_Climate.pdf.

NPS, Inventory and Monitoring Program (Sonoran Desert). "Climate Change in the Sonoran Desert." Accessed July 10, 2018. https://www.nps.gov/articles/climate-change-in-the-sonoran-desert.htm?utm_source=article&utm_medium=website&utm_campaign=experience_more.

NPS, Inventory and Monitoring Program (Southern Colorado Plateau). "Climate Change in the Southwest: Potential Impacts." Accessed July 10, 2018. https://www.nps.gov/articles/series.htm?id=14AD164D-A7DB-C188–37B4B79 60336767E.

———. "Climate Change on the Southern Colorado Plateau." https://www.nps.gov/articles/southern-colorado-climate-change.htm, accessed July 9, 2018.

——. "Series: Climate Change in the Southwest." Accessed June 2018. https://www.nps.gov/articles/series.htm?id=14AD164D-A7DB-C188–37B4B796033 6767.

NPS, North Cascades National Park. "Climate Change Resource Brief." January 30, 2018. Accessed March 22, 2018. https://www.nps.gov/noca/learn/nature/climate-change-resource-brief.htm.

NPS, Olympic National Park. "Animals," Fire History," "Glaciers and Climate Change," "Subalpine Forests," and "Temperate Rain Forests." Accessed April–May 2018. https://www.nps.gov/olym/learn/nature/.

NPS, Sequoia National Park. "The General Sherman Tree," "Geology Overview," and "Plants." Accessed July–September 2019. https://www.nps.gov/seki/learn/nature/.

NPS, Yellowstone National Park. "1988 Fires," "Amphibians," "Canada Lynx," "Geology," "Mammals," "Moose," "Nature and Science," "Plants," "Research," "Wolverine," and "Yellowstone Lake." Accessed February–May 2019. https://www.nps.gov/yell/learn/nature/index.htm.

NPS, Yosemite National Park. "Natural Resource Statistics" and "Snow Creek Cabin." Accessed July–November, 2019. https://www.nps.gov/yose/.
Oregon Historical Society. "The Oregon Encyclopedia: Tillamook Burn." Accessed June 11, 2021. https://oregonencyclopedia.org/articles/tillamook_burn/#.WuovDYjwZjc.
Patterson, Brittany. "Should Iconic Lake Powell Be Drained?" *Scientific American E&E News*, October 27, 2017. Accessed August 12, 2018. https://www.scientificamerican.com/article/should-iconic-lake-powell-be-drained/.
Portland State University. "Glaciers of Wyoming." Accessed March 31, 2019. http://glaciers.us/Glaciers-Wyoming.html.
Public Policy Institute of California. "Dams in California." Accessed October 18, 2019. https://ppic.org/publication/dams-in-california/.
Sierra Club. "David Brower (1912–2000): Grand Canyon Battle Ads." Accessed July 6, 2018. https://content.sierraclub.org/brower/grand-canyon-ads.
Skelton, George. "Capitol Journal: California Should Stop Thinking about More Dams; The State Is Brimming with Them." *Los Angeles Times*, March 4, 2019. Accessed June 11, 2021. https://www.latimes.com/politics/la-pol-sac-skelton-water-storage-california-20190304-story.html.
Sottile, Chiara. "Navajo Nation Votes Down Controversial Hotel and Tram Project at Grand Canyon." *NBC News*, November 1, 2017. Accessed July 2, 2018. https://www.nbcnews.com/news/us-news/navajo-nation-votes-down-controversial-hotel-tram-project-grand-canyon-n81666.
Turner, Monica. "Climate Change and Fire: What Are the Models Telling Us?" PowerPoint presentation. Northern Rockies Fire Science Network. October 2018. Accessed May 4, 2019. https://www.nrfirescience.org/sites/default/files/3_Turner_Part1_ClimateModels_Oct2018.pdf.
Union of Concerned Scientists. "Lake Mead, NV/AZ, USA." Accessed August 12, 2018. https://www.climatehotmap.org/global-warming-locations/lake-mead-nv-az-usa.html.
University Corporation for Atmospheric Research (UCAR) Center for Science Education. "Climate Modeling." Accessed July 26, 2018. https://scied.ucar.edu/longcontent/climate-modeling.
US Bureau of Reclamation Mid-Pacific Region. "Central Valley Project." Accessed October 18, 2019. https://www.usbr.gov/mp/cvp.
USFS. "Survey Finds 18 Million Trees Died in California in 2018." Press release, February 11, 2019, with accompanying data sheets. Accessed November 6, 2019. https://www.fs.usda.gov/detail/catreemortality/toolkit/?cid=FSEPRD60912.
USGS. "FAQ on Invasive Lake Trout in Yellowstone Lake." Accessed May 31, 2019. https://www.usgs.gov/centers/norock/science/faq-invasive-lake-trout-yellowstone-lake?qt-science_center_objects=0#qt-science_center_objects.

———. “Glacier Monitoring Studies.” Accessed September 13, 2018. https://www.usgs.gov/centers/norock/science/glacier-monitoring-studies?qt-science_center_objects=0#qt-science_center_objects.
———. “Repeat Photography Project.” Accessed September, 2018. https://www.usgs.gov/centers/norock/science/repeat-photography-project?qt-science_center_objects=0#qt-science_center_objects.
———. “Retreat of Glaciers in Glacier National Park.” Accessed March 22, 2018. https://www.usgs.gov/centers/norock/science/retreat-glaciers-glacier-national-park?qt-science_center_objects=0#qt-science_center_objects.
Water-data.com. Accessed March 14, 2018. http://lakepowell.water-data.com/index2.php.
Webb, Bob, and Chris Magirl. “The Changing Rapids of Grand Canyon—Crystal Rapid.” Accessed July 14, 2018. https://www.gcrg.org/bqr/15–2/crystal.html.

INDEX

Illustrations are denoted by *p* and the plate number.